How Water Influences Our Lives
水如何影响我们的生活

[挪威] 珀·雅润（Per Jahren）
[中国] 隋同波（Tongbo Sui） 著

How Water Influences Our Lives
水如何影响我们的生活

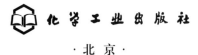
·北京·

作者从地理环境、科技人文、生活万象与可持续发展等多个方面将水这一人类消费量最大的商品的千姿百态展现给读者，生动地展示出了水如此奇妙而耐人寻味的物质特征，既简单又复杂，既司空见惯又奇特新颖，既一目了然又难以预测，既柔软平和又无坚不摧。虽然水属于我们，但却并未能真正为我们所分享……

作者用大量、翔实的数据及全球旅行中拍摄的照片，以自己的亲身经历和体会，阐述水如何影响了我们的思想、文化和生活，如何影响我们的环境，如何成就交通运输，如何产生和吸收能量，如何在全球的动态分布及变化等，以期启发读者对于水的全方位的认识、思考和反思，唤起人类对于水资源的珍视、爱惜及可持续开发应用。

全书共分11章，包括200多篇参考文献。分别从水的综论、水与文化、雪与冰、海洋、河流与湖泊、水作为交通运输之源、水能源资源、水的冲刷侵蚀、水的消耗、水与休闲娱乐、水与食物来源等众多方面阐述了作者的独特观点，阐述了水对生活的影响。

本书可供对水和环境感兴趣的人士，以及对水综合利用感兴趣的人士阅读。

图书在版编目（CIP）数据

水如何影响我们的生活= How Water Influences Our Lives：英文/（挪威）珀·雅润，隋同波著. —北京：化学工业出版社，2020.1
　ISBN 978-7-122-27520-2

Ⅰ.①水… Ⅱ.①珀… ②隋… Ⅲ.①水-影响-生活-普及读物-英文 Ⅳ.①TS976.3-49

中国版本图书馆CIP数据核字（2018）第200993号

本书由化学工业出版社与Springer出版公司合作出版。版权由化学工业出版社所有。本版本仅限在中国内地（大陆）销售，不得销往中国台湾地区和中国香港、澳门特别行政区。

责任编辑：吴　刚　　　　　　　　　　　　封面设计：关　飞
责任校对：宋　玮

出版发行：化学工业出版社（北京市东城区青年湖南街13号　邮政编码100011）
印　　装：北京新华印刷有限公司
880mm×1230mm　1/16　印张18½　字数945千字　2020年6月北京第1版第1次印刷

购书咨询：010-64518888　　　　　　　　　售后服务：010-64518899
网　　址：http://www.cip.com.cn
凡购买本书，如有缺损质量问题，本社销售中心负责调换。

定　　价：298.00元　　　　　　　　　　　　　　　　　　　　　版权所有　违者必究

Preface

Water is the life giving and the origin of all matters. This is not just a myth.
The fact is that:

- Water is a source of human being's culture and civilization;
- Water is in us and around us in an endless buffer. More than 70 % of the surface of the earth is water, and human body generally contains 70 % of water;
- There is enough water on the earth;—still there is a lack of water;
- Philosophically or culturally water is a phase, a process, a movement, a virtue, and a status.

As summarized in Tao Te Ching by Lao Zi, "The highest level of ethics is like water. Water knows how to benefit all things in the world but never contends its own contribution or competes with them. It stays in places loathed by all men, it therefore comes near the Tao".
Water has many folds of characters:

- Water is inclusive and philosophical;
- Water can temper force with grace and mercy;
- Water has very strong cohesion and infinite capacity;
- Water is clear, transparent, fair, and perfectly open in all one's actions;
- Water has the "power" of two opposite sides—blessing and catastrophic.

Few things are like water—a subtle and thought-provoking substance, being so simple yet complicated, so obvious yet unpredictable, so soft and peaceful yet powerful, and so unexceptional yet special and even unique. Water is by far the world's biggest commodity. It is everyone's property, but still it is not obvious for everyone.

Still, and may be just because of this, we very seldom make time to wonder about the many folds and how water in nearly unbelievingly many ways affects our life.

This is what this book is about:

- Naturally the properties of water and the many phases of water from ice to gas, as well as the many faces of water in Chap. 1;
- Philosophically or culturally how water influences our language, our society, our history, and human beings ourselves in Chap. 2;
- Geologically the increasable resources of water on the earth on how it is distributed in snow and ice, oceans, rivers, lakes, etc. And functionally how these water resources influence our life and our world, as in Chaps. 3–5;
- Materially how water creates transport possibilities (Chap. 6), how water gives and takes energy (Chap. 7), how water constantly is changing the surface of the earth (Chap. 8), how it is enough water on the earth, but still it sometimes gets too much or too little (Chap. 9), as well as how water gives us recreation experiences (Chap. 10), and how water supply us with food in a myriad of many folds (Chaps. 11).

It would be too ambitious to believe that one could manage to cover all the ways of water, but hopefully the book, through showcasing the versatility of water for human culture and activities as well as sharing the outcome of the trips to many places of the world and what have been thought and enlightened in mind by the two authors, might contribute to increased interest and insight in the most important substance in our life and society in the past, present and future—water for the researchers, students, decision makers, and everyone who are interested in water-related science and culture.

This edition is co-published with Springer. In accordance with Springer's edition, this book follows the typeset of Springer, including, but not limited to, fonts, size, subscript, superscript, normal or italic letters, as a courtesy.

Asker, Norway   Per Jahren
Beijing, China   Tongbo Sui
September 2014

Contents

1 **Water** 1
    The Phases 7
    References 18

2 **Water and Culture** 19
    References 37

3 **Snow and Ice** 39
    Ice Industry and Ice Export 47
    References 57

4 **The Oceans** 59
    The Channels 66
        The Suez-Canal 66
        The Panama Canal 67
    Sizeable (Water)—Rich Island Nations 68
    The Reefs 69
    The Fiords 72
    Ocean Interests 74
    References 77

5 **Rivers and Lakes, and a Bit More** 79
    Lakes 97
    Moors, Wet Areas, and Swamps 104
    References 106

6 **Transport Lanes** 109
    References 137

7 **The Energy Source** 139
    Geothermic Energy 143
    Waterfall - Energy 147
    Small Power Stations 156
    Other Water Power Solutions 156
    Tidal Wave Electrical Power 157
    Wave Power 157
    Ocean Heat Power 158
    Ocean Current Energy 158
    Water—The Giant Energy Consumer 159
    References 159

8 **Erosion** 161
    Rain and Melting Erosion 165
    River-Erosion 166
    Wave-Erosion 166
    Glacier Erosion 168
    Ground Water Erosion 169

	Freeze-Thaw Erosion	176
	Also Human Created Structures Are Attacked	177
	References	178
9	**Water Consumption**	179
	Desalination	185
	Lack of Water	187
	References	195
10	**Recreation**	197
	The Fountains	197
	River Cruise	205
	Spa	208
	Swimming	220
	Scuba Diving	229
	Golf	231
	References	250
11	**The Food Source**	251
	A Simple, "Quick" Fish-Dish	271
	Artificial Fish Reefs	277
	References	285

Water 1

Water, water
Water in your mouth
Water over your head
Head over the water
Water on the mill
To be mirrored in the water
Water mirror
Jump into the water
In deep water
To keep floating
Walk on water
Water border
Baptizing water
Baptizing font
Fountains
High tide
Tidal water
Water level
Water mist
Water colors
They made waves
Flood
Alike as two drops of water
Dew drops on leaves
Dew fresh morning
Clear as water
When it rain on the pries, it drips on the clergyman
Rain in the hair
Dripping from the roof…
Wet
Brook tingling
River whistling
Waterfall thunder
Many small brooks make a large river
Is the blue Danube brown or muddy?
Crystal clear water
The bluest blue
As long as you have water in your pipe
Like peeing in the ocean
Much water has run into the sea….
Eyes like bottomless lakes
Rain summer
Rainbow
Snow winter
Snow melting
Silk snow
Water and bread
Sky pump
Water pump
Water cock
Water pipe
Water post
The water of life
Well water
Low tide
Water

According the myths—water is the life giving, and the origin of all matters (Figs. 1.1, 1.2, 1.3, 1.4 and 1.5).

In the old Nordic mythology described in the book Edda written by the Icelandic author Snorre Sturlason [1], who tells about the beginning—the creation:

> That the blood from the Good Yme made up the lakes and the oceans, while the earth was created from his flesh, the mountains from his bones.

A flood or deluge myth is a common and widespread theme of all ethnic groups in the world. The basic structure of the nation's flood myths is similar though there are many different places of specific details. Myth arises from the original clan society, which is the process of "imagination or with imagination to conquer the forces of nature, to control the natural forces and to symbolize the forces of nature," as said by Karl Marx, the famous German philosopher [2]. Particularly in China with 56 nationalities or ethnic groups, there are close to 400 legends on flood myth. More will be detailed in Chap. 2.

From ancient time in the fifth century B.C. and up to nearly 250 years ago, water was regarded as one of the four

Fig. 1.3 "Walk on water"—a sculpture at Aker Brygge, Oslo—a winter morning

Fig. 1.1 Spring brook, Lake Sem, Asker, Norway

Fig. 1.2 Water in the mouth—experiences might be many and in more than one sense

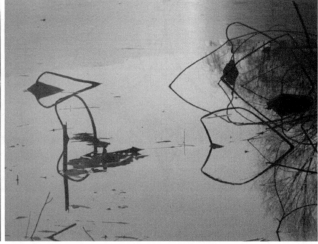

Fig. 1.4 "Water as mirrors"—in a lake in north suburb of Beijing, China, where the partly frozen water created more beautiful views at different times of a day in December 2012. *Left* 15:30 pm, *Right* 16:30 pm

Fig. 1.5 "Water as mirrors"—Buildings and their reflections stand in between the sky above and in the water, in June 2013, Trondheim, Norway

basic elements: fire, air, water, and soil. This can be shown as (Fig. 1.6):

The key concept of the Four Elements raised by Empedocles, a Greek philosopher, scientist, and healer, is that all matter is comprised of four "roots" or elements of earth, air, fire, and water. These elements are not only as physical manifestations or material substances, but also as spiritual essences. He associated these elements with four Greek gods and goddesses—air with Zeus, earth with Hera, fire with Hades, and water with Nestis. The interaction of the four elements is influenced by the relationship between the two great life energies of Love and Strife [4].

This seems similarly yet different from the The Wu Xing (五行 wǔ xíng), including Wood (木 mù), Fire (火 huǒ), Earth (土 tǔ), Metal (金 jīn), and Water (水 shuǐ), also known as Five Phases, the Five Agents, the Five Movements, Five Processes, Five Virtues, or the Five Steps/Stages. This is a fivefold conceptual scheme that many traditional Chinese used to explain a wide array of phenomena, from cosmic cycles to the interaction between internal organs, and from the succession of political regimes to the properties of medicinal drugs. The translation of Chinese Wu Xing as the Five Elements, according to Nathan Sivin's opinion was a false analogy because the classical Greek elements were concerned with substances or natural qualities,

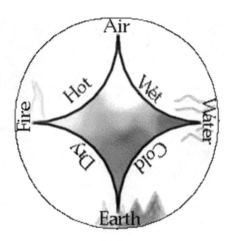

Fig. 1.6 Image from the Particle Adventure Home Page, http://www-pdg.lbl.gov/cpep/four_elem_ans.html [3], in which the matter "sphere" contains four elements and will be unified by love and disintegrated by strife [4]

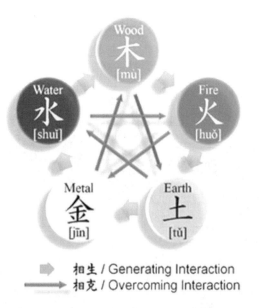

Fig. 1.7 Diagram of the interactions between the Wu Xing. The "generative" cycle is illustrated by white arrows running clockwise on the outside of the circle, while the "destructive" or "conquering" cycle is represented by *red arrows inside the circle* [5]

while the Chinese Five Xings are primarily concerned with process and change in essence [5] (Fig. 1.7).

As shown in the diagram, the order of "Wood, Fire, Earth, Metal and Water" is known as the "mutual generation" (相生 [xiāng shēng]) sequence. In the order of "Wood, Earth, Water, Fire, and Metal " they are mutual conquest" (相胜 [xiāng shèng]) or "mutual overcoming" (相剋 [xiāng kè]) sequence. The system of five phases was used for describing interactions and relationships between phenomena. After it came to maturity in the second or first century B.C. during the Han dynasty, this system was employed in many fields of early Chinese thought, including seemingly disparate fields such as geomancy or Feng shui, astrology, traditional Chinese medicine, music, military strategy, and martial arts [5].

Water is life, water has for millions of years formed our globe, water is the most important substance for human life, and for the animal and plant life around us, and for our culture and our economy.

Water, together with the sun, is the bases for climatic evaluation and possible climate changes. The floating water makes our planet special. No other planet has floating water at least in our solar system. More than 70 % of the surface of the earth is water, and the human body contains 70 % of water.

Water is in us and around us in an endless buffer. Even small changes in the water supply will mean considerable changes to the societies concerned as well as to our bodies and life.

In the introduction to the annual book about weather in 2005, Siri M. Kalvig claims that the weather also means conversation topic, enthusiasm, and politic. Extending this to water does not make it less true (Fig. 1.8).

An Internet—search on *"water"* in September 2012 gave 230 million hits in 0.18 s.

Few items are so simple, and still so complicated as water. The specialists, the hydrologists (*hydrology = the teaching about water*) can answer many questions, but still they cannot foresee where the next bubble comes in the boiling water kettle on the oven. The meteorologists tell us about the weather based on information from the most advanced and modern equipment and communication that can be thought of, but still their aiming efficiency is far from 100 % correct (Fig. 1.9).

Water covers 71 % of our globe, with an average depth of well over 3.7 km. Most of this water is salt water. We call our planet "Planet Earth." More than one have said that the name rather should be "Planet Water." However, the water is rather the wardrobe that protects the life on planet earth during changing conditions.

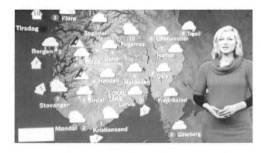

Fig. 1.8 The weather forecast is the most important news item for many

1 Water

Fig. 1.9 Many things in life is predictable, but few things is as unpredictable as where the bubbles come when we are boiling water in the oven

Water is, without comparison, the biggest commodity on earth. Water belongs to everyone. Still, ownership to water has caused strong arguments, quarrels among neighbors, fall of politicians, and armed conflicts. It is claimed that it has been more than 4000 years since water was a direct cause for war. At that time it was in Mesopotamia, however, that water has not only been an underlying cause for conflicts, but also the reason why neighboring states with generally bad relationship have gone to the negotiation table to try to solve the water question.

There is enough water on the earth, still there is a lack of water.

Lack of water, or rather the right and clean water, is definitely the biggest health problem in the world.

Bangladesh is one of the countries with most critical lack of water. However, and still in 1988 the country experienced that 62 % of the country's 144 000 km^2 was put under water by flooding of the rivers Ganges and Brahmaputra.

The anticipated climate changes mean both more lack of water and problems with too much water, and both reduced crops in agriculture in some areas and increased in others. Change in the distribution of water changes the world (Figs. 1.10, 1.11, 1.12 and 1.13).

More than half of the population in the world lives along the coast, in river deltas, or along rivers and lakes. Therefore, it is not strange that water catastrophes are the ones that cause the biggest problem, take most lives and cost the most.

Fig. 1.10 Water on the earth

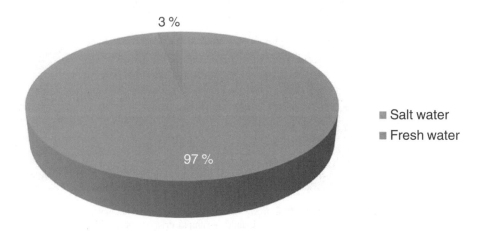

Fig. 1.11 Fresh water on the earth

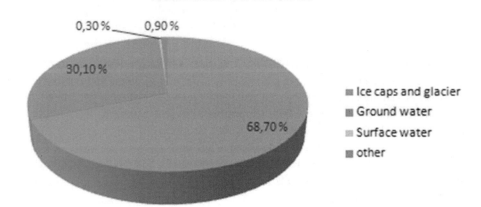

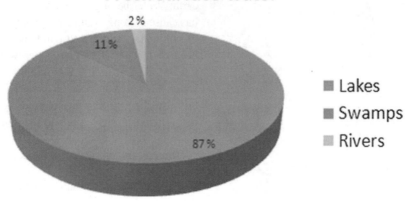

Fig. 1.12 Surface water on the earth

Fig. 1.13 Swamps make up 11 % of the fresh surface water on the earth. If the water in the moors is registered under swamps, or as ground water, is somewhat undecided, and is also a matter of definition. According to Wikipedia [6], a moor is not a swamp, because of the growth on it. The moors are developed when the dewatering is small and at the same time the precipitation is bigger than the evaporation. That is why moors are mostly found in the cooler and tempered area of the world. The largest moor areas are found in the world's two largest countries; Russia and Canada, each with at least 1.5 million square kilometers of moors. In Norway, there are about 30 000 km^2 of moor areas. The picture is from an Asker moor, Norway, on a beautiful Saturday in October

Fresh surface water makes up: 0.3 % of 3 %, or 0.009 % of the world's water resources.

Water is a strange and marvelous molecule that it expands when frozen and boils at far too high temperature. But if it acted in any other way, life as we know would be impossible. A peculiarity of water is that it becomes lighter as it freezes rather than becoming denser like most other substances. This odd property prevents lakes and ponds from freezing into solid during winter.

Water is a chemical combination of two hydrogen atoms and one oxygen atom in each of its molecules. Most often it is formed by a reaction between hydrogen and oxygen.

$$2H_2 + O_2 \rightarrow 2H_2O$$

There are also many other chemical reactions that lead to the formation of water (Fig. 1.14).

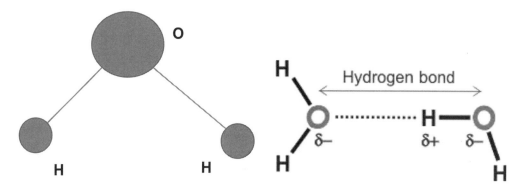

Fig. 1.14 Water molecule (*left*) and illustrative hydrogen bonding (*right*) [7]

In liquid water, the mean O–H length is about 0.097 nm, the mean H–O–H angle is about 106°. Besides, in view of Periodic Table of the Elements, we will find it not only has a smaller volume, but much lighter molecular weight (18) as well compared with the four other common atmospheric molecules, oxygen (32 for O_2), nitrogen (28 for N_2), argon (40 for Ar), and carbon dioxide (44 for CO_2). As a result, both liquid and solid water (ice) have high densities of molecules [7].

Water properties in general:

• Freezing point	0 °C
• Density of ice at 0 °C	0.92 g/cm³
• Density of water at 0 °C	1.00 g/cm³
• Boiling point	100 °C
• Heat of evaporation	600 cal/g

Water has a much higher electrical conductivity than other liquids at room temperature. The current is transferred by ions by the breaking up of water molecule:

$$H_2O \leftrightarrow H^+ + OH^-$$

Based on this property of water, electrolysis of water is conducted as a chemical process to give oxygen and hydrogen again, for which the reaction is indicated as

$$2H_2O + energy \rightarrow O_2 + 2H_2$$

In industrial process considering the fact that pure water, due to its higher electrical conductivity, is difficult or slow to electrolyze, electrolytes are normally included in the solution in order to increase the electrolysis efficiency.

The most frequently used temperature scale, Celsius scale (°C) is based on the behavior of water, where 0 °C is the freezing point, and 100 °C is the boiling point of water. The technical temperature scale, Kelvin (developed by the British physicist and engineer William Thomson Kelvin) is a temperature scale that follows the steps in the Celsius scale, but where 0 K is the absolute zero point at—273.15 °C. The freezing point in the scale then becomes 273 K and the boiling point 373 K.

Water has its highest density at 3.98 °C. At this temperature, the water has a density of 1.00000 g/cm³. At 0 °C the density has sunk 0.99987 g/cm³. This is one of the reasons why water in deep lakes does not freeze that easy in the winter. When cooling to 4 °C heavier water will sink to the bottom, while lighter and warmer water rises to the surface. When water freezes to ice, the density of the ice at 0 °C is lower as 0.916748 g/cm³. The ice is therefore lighter than water and is floating on the surface.

The Phases

Water as is well known at atmospheric pressure exists generally in three phases—solid, liquid, and gas (vapor). The inter-transformation of the three phases is shown below, which can be easily observed in daily life. In order for water to change from a solid to liquid and finally to a gas, the water molecule must gain energy, while the reverse process will release heat. The energy absorbed by water is used to break the hydrogen bonds between groups of molecules.

Besides the well-known principle on water phases, a recent new book written by Professor Gerald H Pollack revealed the fourth phase of water [8]. He gave the explanation as below:

> The "exclusion zone" (EZ), the unexpectedly large zone of water that forms next to many submersed (hydrophilic) materials, got its name because it excludes practically everything. The EZ contains a lot of charge, and its character differs from that of bulk water. Sometimes it is referred to as water's fourth phase.

Fig. 1.15 The hydrological cycle [9]

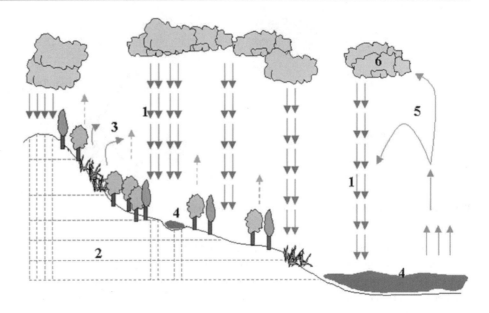

Interfacial zone (IZ) of water or water solution has been ubiquitously studied and can be explained by electric double layer theory in electrochemistry. For instance there might be some amount of siliceous ion groups from glass container which makes some difference. Yet the largest difference may come from interaction between the electrical charge of the glass interface and polarity of water molecules, which makes the characteristics of IZ water differ from bulk water. It is therefore very controversial to characterize water in IZ as a separate phase.

Now, we know it is because of the properties of water upon changing phase under different temperatures that maintains a hydrological cycle of the nature. This cycle starts over again and will go on and on (Fig. 1.15).

In which, 1: Rain or Precipitation by clouds in the sky; 2: Infiltration as ground water to form aquifers, through which to sea or ocean; 3: Transpiration of plants excreting ca. 10 % of precipitation as vapor; 4: Surface run-off, part of rainwater that does not infiltrate into the soil goes to rivers and lakes; 5: Evaporation, through which water rises up and back into the atmosphere and forms clouds that will eventually cause rainwater to fall back on earth; 6: Condensation, vapor transforms back to liquid (dew, water in the air or cloud) or solid (snow, ice, frost)

Phase changes of water mean a lot to mankind. Water as we know is one of the lightest gasses yet forms a dense liquid. The volume change is the greatest known as 1603.6 folds at the boiling point and standard atmospheric pressure. This change in volume allows water to be of great use in the steam generation of electrical power. More will be discussed in Chap. 7 on water as energy source (Figs. 1.16, 1.17, 1.18, 1.19, 1.20 and 1.21).

The Liquid phase, which is what we normally associate with water, consists of water molecules, and is connected together with hydrogen bonding. When ice melts, regular lattice arrangement of water molecules disappear and many of the bonds are dissolved, and what is left is not enough to keep the water together in a regular pattern. The maximum density of water occurs around 4 °C. When water is heated from 0 to 4 °C, it contracts, while at higher temperatures, it expands. This phenomenon is connected to the increased reduction in hydrogen bonds (Fig. 1.22).

Water goes in an infinite cycle. Evaporation from the sea, lakes, and rivers is cooled when the gaseous water rises and is compressed and formed into clouds. When the density in the cloud is high enough, it becomes either rain or snow. The precipitation sinks to the ground or to the ocean, runs into brooks, outlets, or rivers or the water evaporates again or forms new clouds. Plants and trees suck up water from the ground through their roots. From the leaves water is evaporated. When the water comes back to the sea, the circle is completed. This cycle is also called the hydrological cycle. The SSB (Statistisk Sentralbyrå = Statistical Central Bureau) in Norway, based on information from the Norwegian Waterway and Energy Directorate, and the Norwegian Meteorological Institute [10], gives the following figures for the Norwegian water balance in the years 1961–1990 (in million m^3) (Fig. 1.23):

Fig. 1.16 Phase change. The spring has arrived, and the water goes directly from a solid phase to a gaseous phase. The old footsteps have a harder consistency and will be the last to melt

Supply:

Precipitation	470 671
Supply from neighbor countries	12 394
• Evaporation	112 000
Recourses	**381 260**
Runoff to the coast (ocean)	369 375
Runoff to neighbor countries etc.	11 885

Sometimes we experience dew inside a window, we find moisture on the outside of a drinking glass, or we find dew on the grass or frost on the ground or the car. The ability of the air to absorb the moisture depends on the temperature. At high temperature the air has the ability to absorb more humidity. At the same time, the relative humidity might be low because the percentage part of gaseous water in relation to the moisture that forms condensation is low. On the opposite, we might experience that the absolute humidity is low while the relative humidity is high.

The dew point is the temperature in the air that the air must be cooled down to at constant air pressure so that the moisture shall condensate to water. This means that the relative humidity at this temperature is 100 %. The dew point represents about 1 g of vapor per kg air at −15 °C and 14.5 g vapor per kg air at 20 °C (Fig. 1.24).

Looking at the figure above, we see that the air at 30 °C can absorb about 30 g per kg at this temperature. If we have air with 20 % relative humidity and fill a glass with ice water, the temperature just on the outside of the glass might be down to 5 °C. The air near the glass is so low that the moisture might condensate even if the general air humidity is low (Figs. 1.25, 1.26 and 1.27).

When the nationally very famous Norwegian author, poet and later war hero Nordahl Grieg, was correspondent in China in the 1930s, he wrote the "homesick-poem": *WATER* (Figs. 1.28 and 1.29)

> In the first 3 verses he describes the hot summer and his life in Shanghai, often with gin and bitter from the bar in his hands.
> In the next versus he describes his homesickness and tells that he soon will go home to Norway.
> Still ordering another drink, - Most of all he is longing for the clear, cold Norwegian water
> He is dreaming of the sound of clear water from brooks and small rivers, water in spring, water in the autumn
> The long poem ends with: Dear Jesus, give me water!

Some water records:

- Most rain in 24 h: 1870 mm, 15–16.03.1952, Chilaos, La Reunion [12]
- Highest average annual precipitation: 11 874 mm, Mawsynram, Meghalaya, India [12].
- Highest measured annual precipitation: 26 461.7 mm from 01.06.1860 to 31.07.1861, Cherrapunji, Meghalaya, India [12].
- Most snow in the ground: 11 455 mm in March 1911, Tamarac, California, USA [12].

Fig. 1.17 Winter of Beidaihe, Hebei Province, north of Bohai Sea where it was very cold with frozen seashore in January 2012. *Photo* Airong Gong. The freezing point of water decreases from 0 °C for pure water to below as salt concentration increases. At typical salinity of seawater about 3.5 % (35 g/L, or 599 mM), an average of world oceans, it freezes at about −2 °C

Fig. 1.18 Unique beauty of rime brings bald tree new life in winter, charm of phase change of water. January 2011, Jilin, China. *Photo* Lei Song

- Highest Norwegian normal annual precipitation: 3 575 mm, Brekke, Sogn og Fjordane [13].
- Highest Norwegian precipitation in 24 h: 229.6 mm, Matre, Hordaland. November 26, 1940 [13].
- The most magnificent tidal bore: Qiantang River Tide, a world wondrous spectacle which is happening around August 18 of Chinese lunar calender with a high tidal range as high as 8.1–9.0 m, northwest of Zhejiang Province, China.
- Largest strait in China: Taiwan Strait with total length of 380 km and average width 190 km.
- Highest river in China: the Yarlung Zangbo [Yalu Tsangpo] River in Tibet with an average 4500 m above the sea level and a length of 2057 km inside China's

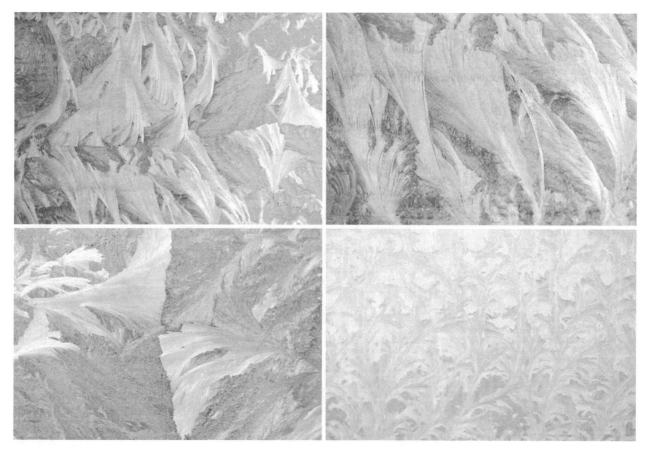

Fig. 1.19 Charm of nature—phase change of water. When moisture inside the warm room meets the cold glass window at winter night, beautiful ice flowers form—a masterpiece of nature and a symbol of celebrating the Spring Festival, the most important Chinese New Year in January 2011 in a coastal town, Rushan in Weihai, Shandong Province, China

Fig. 1.20 Phase change of water can occur like this. The frost deposited and grew along the vein of Chinese cabbage beautified the nearly withered leaf, in an early morning of November 2012, Italian Farm, suburb of Beijing

Fig. 1.21 Charm of phase change of water. Beautiful like this, who can deny? The frost deposited beautified the withered Chinese rose, in an early morning of November 2012, Italian farm, suburb of Beijing

Fig. 1.22 Phase transformation. Spring at Lake Sem in Asker, Norway. Water goes from solid phase to liquid phase

territory. The total length is 3848 km covering China, India, and Bangladesh.

- Largest hail: A dramatic hail weather killed 92 people in Bangladesh on April 14, 1986. The hails had the size of tennis balls and weighed several hundred grams [13].
- Largest snow particle: Siri Kalvig [13], with reference to Guinness Rekordbok claims that on January 28, 1887 in USA a snow particle of 38 cm in width and 8 cm thick was recorded.
- Biggest snow fall: 5.58 m between 13th and 19th of February 1959, Mount Shasta, California, USA [13] (Fig. 1.30).

The Gaseous phase, consists of water molecules that move freely versus each other. Every atom has a size of about 3 nm. In general, the end of the hydrogen atoms of water molecule is positively charged in comparison to oxygen atom—end. When two molecules are close to each

Fig. 1.23 The clouds can sometimes be very low. Here we see the clouds as a wet autumn fog over the Skaugum hill in Asker, Norway. We can imagine the characteristic profile of the hill, but it is continuously in change. Even if we cannot feel the wind, the air currents around the hilltop change the visual views of the formations (The pictures are taken within a time frame 20 min)

Fig. 1.24 The ability of air to absorb humidity depending on the temperature

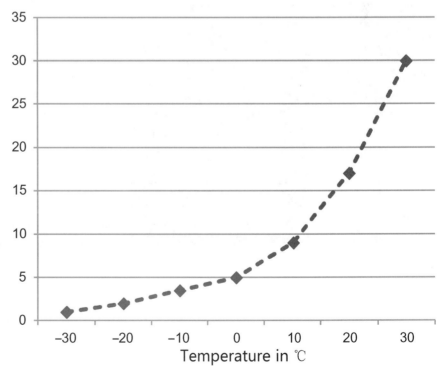

Fig. 1.25 Fresh water meets salt water. The city river meets the sea, Strømstad, Sweden

◀ **Fig. 1.26** Rain!—Rain water is in principle the cleanest water, because it has been through a natural distillation process—but it might contain smaller or larger impurities as dissolved gasses, salt, and dust. The impurities might be salt from the ocean, volcanic dust or other dust, residues from vegetation and industrial emissions. In particular, emission of sulfur dioxide has led to so-called "acid precipitation." Acid precipitation from the European continent has led to acidification of water ways in Southern Norway. As "acid precipitation" is normally considered precipitation with a pH value below 5.6. Sometimes pH values as low as 4 have been recorded in Norway. The lowest pH values from acid precipitation are 1.7 from the USA. This is 10 000 times more acidic than pure rain water [11] (The pH value is 7 when a liquid is neutral. Maximum acidic is 0, and pH of 14 means that the liquid is maximum basic)

other and reasonably oriented, the hydrogen atom in a molecule will be attracted to the oxygen atom in another molecule and create a weak bonding called hydrogen bonding. This hydrogen bonding between neighboring water molecules, together with the high density of molecules due to their small size, produces a great cohesive effect within liquid water that is responsible for water's liquid nature, in particular the life-giving properties at ambient temperature (Figs. 1.31, 1.32, 1.33 and 1.34).

The Solid phase, water has many forms of solid ice depending on the conditions water experiences during the phase change, will normally consist of water molecules bound together with hydrogen bonding in a regular pattern. Probably there are a lot of spaces between the molecules.

Fig. 1.27 Precipitation in snow not only brings water back to the cycle via phase change but also whitens the world. End of November 2012, Moscow, Russia

Fig. 1.28 The famous Geiranger fiord, Norway. What had been the attraction, if it had not been for the water and the water fall?

Fig. 1.29 Water is destructive and creative. Water also contributes to hope and fantasy. *Photo* Espen Jahren

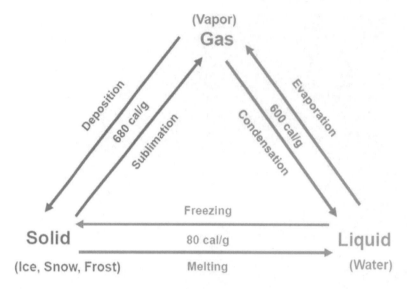

Fig. 1.30 Change of Three Phases of Water

Fig. 1.31 Charm of phase change of water. Hani Rice Terrace, Yuanyang County, Yunnan Province, China. 350 km south of Kunming Changshui International Airport, September, 2012. We can enjoy the beautiful water—mountain scenery with a small village merged in the cloud like fairyland. Hani Terrace was approved by UNESCO in the World Heritage List in June 2013

Fig. 1.32 Charm of phase change of water. Cloud can be so nice. *Left* a white dragon-shaped cloud landed on the Table Mountain; *Right* A colorful dusk above the sea, Cape Town, South Africa, September 2012

Fig. 1.33 Phase change, from the Geysir area in Iceland. The Geysir "Little Stokkur" is blowing. The Geysirs in Iceland have given name to this spectacular phenomena in the nature and visual experience of heat from the ground. Water seeping down into cracks in the underground hits rock that is so warm that the water starts to boil. The water is pressed upwards, and under the pressure difference some of the water is transformed into a gaseous phase

Fig. 1.34 Phase change. Even on a winter day in January with a reasonable amount of snow in the tracks, the organizers of the traditional Holmenkollen Skiing Festival (normally taking place in the first week of March) in Oslo, Norway take no chances. The snow cannons are running on full speed and transforming water to more snow

References

1. Sturlason S.: *Edda*. Gudrun Publishing AS, Island 2003.
2. Marx and Engels, Literary Arts, Vol. 01, Beijing: People's Literature Publishing House, pp. 93 (in Chinese), 1983.
3. Access on May 20, 2014, http://www-pdg.lbl.gov/cpep/four_elem_ans.html.
4. Nathan Sivin, "Science and Medicine in Chinese History," in his Science in Ancient China (Aldershot, England: Variorum), text VII, p. 179. 1990.
5. Access on May 20, 2014, Wikipedia: Wu Xing: http://en.wikipedia.org/wiki/Five_phases.
6. Internet 07.10.12: Wikipedia: *Sump*, http://en.wikipedia.org/wiki/Sump.
7. Martin Chaplin, Hydrogen bonding in water: Introduction, in Book: Water Structure and Science, http://www1.lsbu.ac.uk/water/index2.html.
8. Gerald H Pollack, THE FOURTH PHASE OF WATER, EBNER & SONS PUBLISHERS, SEATTLE WA, USA. ISBN: 978 – 0 – 9626895, pp. 25, 2013.
9. Access on May 20, 2014, http://www.lenntech.com/hydrological-cycle.htm.
10. Statistisk Sentralbyrå (Statistical Central Bureau): *Statistiske analyser(Statistical analysis) – Naturresurser og miljø (Natural resourses and environment) 2005,* Statistisk Sentralbyrå, Oslo-Kongsvinger, Norway, December 2005.
11. Kunnskapsforlaget/ Norges Naturvernforbund: *Natur- og miljøleksikon,* Hovedredaktør: Henning Even Larsen, Oslo,Norway, 1991.
12. Buckley Bruce, Hopkins J. Edward, Whitaker Richard: *Været*, Gyldendal Fakta, Oslo, Norway, 2004.
13. Kalvig Siri M.: *Værårboken (the weather year book) 2005,* Schibsted Forlagene AS, Oslo, Norway, 2004.

Water and Culture 2

As attributed by Thales of Miletus (634–546 BC), water is the source of all life. Similarly in ancient China, Guanzhong (720–645 BC), a legalist chancellor and reformer of the State of Qi during the Spring and Autumn Period of Chinese history, realized more than 2000 years ago this important viewpoint that Water is the origin of all thing. In his famous work "Guanzi Water Place", he proposed that "Water is the intrinsic nature and the source of life," and maintained the implementation of education to water. Water is therefore a source of national culture and civilization.

Study from archaeology, history, and anthropology has proved that most of the human beings civilization originated from water sides. The four ancient civilization countries were resided by rivers—the ancient Egyptian civilization was reward of the Nile, the ancient civilization of Babylon was originated from Euphrates and Tigris in Mesopotamia, the ancient Indian civilization was developed along the Ganges, the sacred river of India, and Indus, and the Chinese civilization of 5000 years began and flourished in the valley regions of the Yellow river and the Yangtze river.

As Ernst CASSIRER said, "From the history that any great culture is controlled and dominated by the flood myth." The mythological stories around water are also many folds, and gods and goddesses with power over or with connections to water are found in all mythologies. There is important similarity in the basic structure on the creation or recreation of the world and human beings. A hero or Heroes appeared and strived to conquer the disaster and rebuilt a new civilization. Nevertheless, different nations have their specific versions or characteristics of the flood myth, and even manifest their different national spirits and cultural values. As an example, the western flood myths are usually God-centered or theocentrism and embody a strong sense of religion with the theme of asylum, while Chinese flood myths are human-oriented and take "harnessing water" as a theme to reflect the use and control of flood.

Of the more well-known water characters that are more or less known by most people even today, we just mention (Fig. 2.1):

- The Bible tells that Noah was the tenth generation after Adam, and he was ordered to build an Arch and to collect two samples, one male and one female of all animals to survive after the flood of sins, where God had the intention to clean the world. After 40 days and 40 nights of rain, the water was covering the whole world, and the Arch stranded on the mountain of Ararat. The mythology in the story is underlined in the claim that Noah should have lived until he became 900 years. The rainbow was the sign that God should never again destroy the earth. A similar story is found in the old Babylonian legends, and possibly is the origin of the legends a major flood in the area between 3 000 and 2 000 BC.
- Poseidon, the God of the sea from Greek mythology, and Neptune from Roman mythology, were mighty Gods that ruled over oceans, water, storms, earthquakes and floods. Nowadays we find the names in a number of relational areas, from special products to harbor bars and cafés (Fig. 2.2).
- The anecdotes from Greek mythology about the three *Sirens* that fooled seamen into death with their songs have been copied in a number of varieties, and the name has even become a word for alarm signals. The mermaids have been a sex-related attraction object in a number of stories and anecdotes (Fig. 2.3).
- The story about the virgin at the Lorelei cliff in the River Rhine from German mythology that fooled the seamen has also become a traveling story. This story has strong bonds to the Nix stories to be mentioned later in the chapter.
- Chinese flood myths are manifold with close to 400 stories among the 56 different national "nationalities." Chinese history books regard the Huang Di, Yan Di, and Yi tribes as three tribal groups 5000 years ago. The former two groups lived in the valley of the Yellow River, and Yi tribes occupied East China.

The best well known and praised flood myth, according to Shan Hai Ching (The Classics books before the Qin

Fig. 2.1 The rainbow has given the apartment building its glory in Asker, Norway in September

Fig. 2.2 "Poseidon" is a brand name in diving equipment

dynasty, mainly describes ancient myths, geography, animals, plants, minerals, witchcraft, religion, history, medicines, folks, ethnic, and other aspects) originated from the control of flood by the great father Gun (who is said as grandson of Huang Di) and his son Yu. Gun built dikes to prevent flood. Yet he failed due to the severe flood. He heard that there was a special soil named Xi Rang by which could stop the flood. He took the special soil from the heaven without the permission of the Emperor of Heaven in order to save the people in suffering. The emperor was enraged and killed Gun. Though Gun had been dead for 3 years, his body had not decayed and flew out a dual-horn dragon when the body was cut open. That was said his son Yu who carried on his un-completed mission of flood control. Yu, having learned from the previous failures and investigated the flood characteristics and topography, dredged waterways and irrigation canals and finally succeeded in conquering flood and benefiting the agriculture of many tribes spread far and wide. Yu was so devoted to his work that he did not visit his home for 13 years although he traveled nearby three times. His celebrated contributions won him the respect of people who honored him as "Yu the Great" and God of the soil.

Another myth worth mentioning is about Fuxi and Nüwa from Miao and other tribes which lived along the Yangtze River valley. Fuxi is said the first man who used ropes to make nets for hunting and fishing. In the days of Nüwa, the four pillars supporting the heaven collapsed and the earth cracked with flame spreading wildly, torrential water flooded everywhere and fierce beasts and birds preying on men. She smelted rocks and made five-colored stones to patch up the

Fig. 2.3 The little mermaid at Langeline in Copenhagen, Denmark is an attractive sightseeing object. The statue is made by the artist Edvard Eriksen, and celebrated in 2013 its 100 year's anniversary, and is still a reminder of the fairy tales from H.C. Andersen with the same name. *Photo* Marit Myhre

subsequently. The history of using pictographs in China can be traced back to 4500–6400 years ago, which was discovered in Shandong province as an evidence of late Da Wen Kou Culture. The unearthed pottery wine vessels from ancient tombs were found to have characters stylizing pictures of certain physical objects. They are therefore named pictographs. For example, water was written as ⟫ stylizing a flowing water, the sun as ⊖, and the moon as ⟩. These pictographs are easy to understand and are close both in style and structure to the inscriptions on the oracle bones and shells, though they antedate the latter by more than 1000 years.

The evolution of water character is given below from ancient oracle to modern regular script:

http://www.zdic.net/z/1c/js/6C34.htm [2]

This represents a change of water character from Oracle to Bronze inscription, Small seal script and Regular script. Interesting is that from the evolution radical 氵 is derived to stylize things or status of things related to water, which can be very popularly found in Chinese characters, for example:

River as 河 [hé] or 江 [jiāng]; Sea as 海 [hǎi]; Wave as 波涛 [bō tāo]; Lake as 湖 [hú]; Wet or Humid as 潮湿 [cháo shī]; Flow as 流 [liú]; Steam as 汽 [qì], etc.

The word Han as 汉 [hàn] (the Han nationality accounts for 93.3 % of Chinese population at present) with water radical to the left is also the best historical witness of Chinese culture in close relation with water.

When water converges it becomes river as ⟫⟫ in oracle stylizing a flowing water, which is somewhat different from the ancient Egyptian water as ∼∼∼, symbolizing a still water with ripples. The difference and common in ancient Chinese and Egyptian on water character and the way of thinking and expressing water might be found in the difference in geological conditions of the rivers between China and Egypt.

The second principle: Associative compounds, to combine two or more elements, where each element has its own meaning, to form a new character expressing a new idea. Ice (冰 [bīng]), for example, in bronze inscription is written as, a pictograph or ideograph of water ⟫ in an expansive solid state. The change of character of ice is shown below from Bronze inscription to small seal script and regular script, which 冫 is derived to stylize cold.

heaven and cut the four legs of a huge turtle to prop up the fallen sky. With water and land restored to order and the fierce animals killed, people again lived in peace and happiness. Nüwa in return was regarded as a goddess for her great achievements [1].

For the especially interested, we might recommend the web page http://en.wikipedia.org/wiki/list_of_water_deities, where you will find examples of water Gods from 26 different mythologies from all parts of the world.

As mentioned earlier, water is a source of national culture and civilization. Water has inspired authors, poets, and other artists through all times in history, both directly in embracement and disgust, in a bit more indirect descriptions and poetic descriptions about the life by, in, and on water.

It might be interesting to start with the evolution of Chinese script—水 (Pinyin: [shuǐ]; English: Water; Latin: Aqua) to introduce water and water-related in Chinese culture.

It is well known that Chinese is a script of ideograms with pronunciation system of Pinyin instead of an alphabetic language. The formation of Chinese characters follows three principles:

Hieroglyphics or the drawing of pictographs—This is the earliest method by which Chinese characters were designed and from which other methods were developed

氵 冰

http://www.zdic.net/z/15/zy/51B0.htm [3]

Same as above, a number of characters can be derived to stylize things or status of things related to coldness, for example:

Cold as 寒 [hán] or 冷 [lěng] or 寒冷 [hán lěng]; Piercingly Cold as 凛冽 [lǐn liè]; Cool as 凉 [liáng] for noun and [liàng] for verb; Dewdrop as 凇 [sōng], etc.

The third: Pictophonetics—to create new characters by combining one element indicating meaning and the other sound or pronunciation. For example, as the above-mentioned Dewdrop as 凇 [sōng], in which the left radical is related to water in cold situation and thus can be condensed either from steam into liquid or solid state or from liquid into solid state, and the right radical represents the pronunciation of the character. The nice pictures about dewdrop can be found in later chapters.

Other Chinese characters related to water might also be interesting to brief:

Steam: Chinese as 汽 [qì], which evolution is shown from Small seal script to Regular script as:

汽

Rain: Chinese as 雨 [yǔ], which evolution is shown from Oracle to Bronze inscription, Small seal script and to Regular script as:

It seems that Rain in oracle is much easier to understand than modern regular script. The pity is that we do not use oracle now.

Snow: Chinese as 雪 [xuě], with rain as the upper radical indicating its origin. The evolution is shown from Small seal script to Regular script as:

Spring: Chinese as 泉 [quán], the evolution shown below from Oracle to Small seal script and to Regular script stylizes water flowing down from a high place:

Pictographic signs have also been used for a number of occasions in modern times. Olympic games have used them to symbolize the different events for some time. During the 29th Beijing Summer Olympic Games in 2008 we found Beijing Olympic Games sports icon, being rewarded as the beauty of the Seal Script, used seal character strokes as the basic form. The formation of the icon combined the pictographic charm of Chinese ancient oracle, bronze inscriptions of with simplified feature of modern design—original creativity, fluent and fruity, tempering force with grace, beautiful and elegant, melodious, which showcase a perfect combination of Chinese culture essence with Olympic sports. It, as sports icon, also delicately fits the requirements of easy to recognize, easy to remember and easy to use.

The following are icons of some of the 35 sports for Beijing Summer Olympic Games in 2008, in which we can easily find all the 8 water-related sports.

Canoe/Kayak Slalom   Swimming   Synchronized Swimming   Diving   Water Polo   Rowing   Canoe/Kayak Flat Water

| Sailing | Baseball | Archery | Judo | Wresling | Shooting | Boxing |

Water has been consistently cited since ancient time in China to symbolize the sublime virtue of a man. This can be seen everywhere, in every dynasty, and are well known in China. We can easily find more than three hundred Chinese idioms that are related to Water if we hit Chinese web search engine BAIDU (http://www.baidu.com). Chinese idioms (中国成语 [zhōng guó chéng yǔ]), as one of the priceless legacies deeply rooted in traditional culture with extremely profound implications in them, have been making Chinese language more powerful, functional and thus more fascinating.

The following are some idioms which can reflect Water's virtue and character.

上善若水 [shàng shàn ruò shuǐ]

One of the most famous idioms coming from Chap. 8 of **Tao Te Ching (Dao De Jing** in Chinese Pinyin or **Classic of the Way and Virtue** in English) by Laozi (also spelled as Lao-Tzu; Lao-Tze, 600 BC), one of most famous ancient Chinese philosopher, poet and legendary figure of Chinese culture. He is best known as the reputed author of the Tao Te Ching and the founder of philosophical Taoism.

This idiom gives the implication that **the highest level of ethics is like water**, which is beneficial for all things, without striving for fame and gain. It implies a philosophy of human beings which seems to be inactive and yet closest to **the Way of nature**.

Water is Inclusive and Philosophical in Chinese eyes, which can be seen from the idiom as:

海纳百川，有容乃大 [hǎi nà bǎi chuān, yǒu róng nǎi dà]

All rivers run into sea, tolerance brings greatness and respect.

This encourages a virtue of being generous, broad-minded, modest, and tolerant to diversity, which eventually leads to a greater success and power, like water, which is always compatible with land and becomes part of land and with life and becomes part of life and never becomes arrogant for these. Finally, it completes itself as mighty sea surging forward.

Water can temper force with grace and mercy. It looks tender and weak and flexible to form any shape to follow any mold, it is on the other side so determined and persistent, and therefore was reputed by Chinese idiom in Song dynasty during the period of 1127–1276:

绳锯木断，水滴石穿 [shéng jù mù duàn, shuǐ dī shí chuān]

It means that "rope can saw off wood, constant dropping will wear away a stone."

It is not for the force but persistent spirit and effort of water that can fulfill the conquering of hardness with softness. Profound morals is brilliantly shining through this and telling us the truth that persistent quantitative change will lead to the final qualitative change.

Confucius (551–479 BC), the most famous ancient Chinese teacher, editor, politician, and philosopher of the Spring and Autumn period of Chinese history also in his masterpiece, **the Analects**, made a subtle comment on water to connote the swift flying of time:

逝者如斯夫，不舍昼夜 [shì zhě rú sī fú, bù shě zhòu yè]

Time is going on like this river flowing away endlessly day and night.

Confucius also made enlightening remarks in **the Analects** as:

智者乐水，仁者乐山 [zhì zhě yào shuǐ, rén zhě yào shān]

The wise loves rivers but the benevolent loves mountains. It indicates that each has his own likings. This actually comes from Confucius in The Analects which said, "The wise enjoy water, the humane enjoy mountains. The wise are active, the humane are quiet. The wise are happy; the humane live long lives."

Water has very Strong Cohesion and Infinite Capacity. Now we know that the cohesive unity comes from the hydrogen bonding between water molecules. It can float ships and do the opposite. Here is the Chinese idiom about this:

水可载舟，亦可覆舟 [shuǐ kě zài zhōu, yì kě fù zhōu]

As the water can float a boat, so can it capsize it.

This proverb actually refers to **the "power" of the water in two opposite sides**, just like the phrase "The same knife cut bread and finger." It is normally used to connote the relationship between people (as water) and ruling government (as ship) and to advise the importance of serving for people.

More examples of idioms related to sailing on water can be given, such as:

逆水行舟，不进则退 [nì shuǐ xíng zhōu, bú jìn zé tuì]

A boat sailing against the current must forge ahead or it will be driven back.

The most common use of this phrase is that study or learning is like sailing against the current: you keep either foregoing ahead or falling behind, which indicate that we have to strive for the work otherwise we will be laid behind since life and work are not easy;

顺水推舟 [shùn shuǐ tuī zhōu]

It means to push a boat in the direction of the current, more specifically to make use of the favorable current situation to push matters through with little effort;

随波逐流 [suí bō zhú liú]

It means to swim or sail with the stream and implies that one must have one's own opinions and not drift with the current.

Water has power and life, which can be expressed in the Chinese idiom:

流水不腐，户枢不蠹 [liú shuǐ bù fǔ, hù shū bù dù]

Running water is never stale, and a door-hinge is never worm-eaten.

This metaphorizes that exercise regularly can have lasting vitality, vigor and vitality.

大浪淘沙 [dà làng táo shā]

It means mighty wave crashing on a sandy shore, which is often used to describe the growing or screening through fierce competition and the nature's evolution through the principle of survival of the fittest.

Some more interesting idioms are as follows:

水中捞月 [shuǐ zhōng lāo yuè]

Catch the moon in the water, which implies it will be just vain and ineffectual effort if one targets on something that is impossible to succeed no matter how hard you work for;

无源之水，无本之木 [wú yuán zhī shuǐ, wú běn zhī mù]

Water without a source, and a tree without roots

It implies a thing without basis, just like a castle in the air without foundation.

饮水思源 [yǐn shuǐ sī yuán]

When one drinks water, one must not forget where it comes from.

One should be gratitude for the source of benefit or remember past kindness.

Similar idiom like:

滴水之恩当涌泉相报 [dī shuǐ zhī ēn, dāng yǒng quán xiāng bào]

A favor in the size of a drop of water should be repaid with the amount of a surging spring.

覆水难收 [fù shuǐ nán shōu]

Spilled water cannot be gathered up.

It's the same meaning as "It is no use crying over spilt milk."

心如止水 [xīn rú zhǐ shuǐ]

One's mind settles or calm down as still water, a state of mind without any distracting thoughts.

柔情似水 [róu qíng sì shuǐ]

Tender and soft as water

This is normally used to describe the love between man and women—their tender love flows like a stream.

Furthermore, water also exhibits virtue and character as follows:

Water can be just perfect. This can be seen through the transformation for the three phases: solid, liquid and gas. Under atmospheric pressure water can be turned into gas (steam) at 100 °C and ice at 0 °C. The change of water's state under different circumstances implies that there exists a limit for everything under certain conditions and that it is important for us to fully understand and take advantage of this property.

Water is clear, transparent, fair and perfectly open in all one's actions. We like to see fish swimming in clear water, we admire water as it can mirror the truth, kindness and beauty of the world. This is because water has a peaceful heart and can be stored in any vessels no matter it is clay vase or golden cup and fit very well any shape of any container.

Now we know how Chinese think and respect water. Here are two idioms as proof:

君子之交淡如水 [jūn zǐ zhī jiāo dàn rú shuǐ]

The friendship between gentlemen appears indifferent but is pure like water.

This is similar to "A hedge between keeps friendship green."

If this becomes too much and exceeds the limit, it then goes to the opposite:

水至清则无鱼 [shuǐ zhì qīng zé wú yú]

When the water is very clear, there will be no fish.

This means that "One should not demand absolute purity, otherwise he who is too critical has few friends."

Now we might understand **Water** at least to some extent through the evolution of Chinese characters and idioms associated with water—**a subtle and thought-provoking substance being so simple yet complicated, so soft and peaceful yet powerful, and so unexceptional yet special and even unique**.

Besides, another spectacular Chinese culture with water is the long history of poems on water. People often use poems to express their mind and thinking, their affection and aspiration, and their success and failure. If we log on the Chinese web search engines on famous ancient Chinese Poems (http://www.gushiwen.org/gushi/xue.aspx [4]) we will see from Xianqin period (2100–221 BC) up to Qing Dynasty (1616–1912) tremendous contributions to Water or related. Poems about Rain rank the first in number with 126, followed by snow with 88, water with 67, mountain and water with 41. We can also find specific number of poems describing Yellow River with 23 and the Yangtze River with 20.

Before that let us start with water to one's daily life. It is known in ancient China that the most luckiest and happiest things in one's life lies in four important moments, among which the first happiness is related to water. The poem can prove this:

Four—Happiness Poem
to have a welcome rain after a long drought,
to run across an old friend in a distant place;
to spend the wedding night,
to succeed in the government examination.

Yes, to have a welcome rain after a long drought, who will not be happy with this?

The first eulogy in Chinese history is 《诗经》 [shī jīng], the Book of Songs, also known as Three Hundred Poems (诗三百 shī sān bǎi), is the earliest collection of poems in China with a history more than 2500 years, recording a total of 305 poems written in 500 years from the early Western Zhou Dynasty (1100–771 BC) to the middle of the Spring and Autumn Period (770–476 BC). It was said that the great sage Confucius made great contribution to it and used it as a textbook to teach his disciples. So the Book of Songs was not only a general collection of ancient Chinese history and literature books, but also conveyed the spirit of Confucianism. Confucius once said, "One could not talk well without learning the Book of Songs." The Book of Songs exerted a very profound effect on ancient China in terms of politics, culture, language, and thinking. Confucius, a sage of China and who gave a high praise to the Book of Songs, claimed that people's cultures, observation abilities, and interpersonal skills could be highly improved through the study of the Book of Songs.

The Reeds and Rushes (part)
When reeds and rushes grew green,
And dews to hoar-frost changed.
One whom they speak of as "that beauty",
Somewhere the river ranged.
Upstream they went in quest of her,
A long and toilsome way.
Downstream they went in quest of her,
In mid-stream there lay!

The Ode to Xizhou (西洲曲 [xī zhōu qǔ]) is the representative work of Folk Songs of the Northern and Southern Dynasties (420–589). The main theme is that the heroine's lover had not come back home, so she went to gather lotus to divert herself from loneliness and boredom. At a distance from the lake, both of them were lovesick apart. Finally, she begged the wind to bring her dream to Xizhou where her lover was staying.

Ode to Xizhou (Extract)
…When they gather lotus at Nantang in autumn,
the lotus blooms are higher than their heads;
They stoop to pick lotus seeds,
Seeds as translucent as water.
…The waters is beyond the scope of eyesight,
You are lovesick, me either…
The south wind knows my mood,
It blows my dream to Xizhou.

A poem with deep significance said like below in cooking peas in the pot:

The story came from the period of Three Kingdoms (220–280) in Chinese history, Cao Pi, the son of King of the Wei State, Cao, came to power. He was said to envy the brilliant talent of his brother Cao Zhi and once requested his brother to complete a poem within seven paces, otherwise he would be executed. The following is said what Cao Zhi did.

Poetizing While Taking Seven Paces by CAO Zhi
Pods burn to cook peas,
Peas weep in the pot:
Grown from the same roots,
Why boil us so hot?

Most of the famous poems related to water happened in Tang Dynasty. Let us enjoy some of them.

Liu Zongyuan (773–819), a well-known statesman and writer in the Tang dynasty and one of the top 8 masters on poetry during the period of Tang and Song Dynasty. The poem below is one of his masterpieces picturing us a solitary but vivid, a rich yet pure imagination.

River Snow by LIU Zongyuan
A thousand mountains without a bird.
Ten thousand paths without trace of man.

(continued)

A lonesome boat and an old man in straw raincoat,

Alone in the snow fishing in the freezing river.

Note A thousand mountains imply all around mountains; Ten thousand paths imply all paths. The number used here is a metaphor only rather than an exact number

Cen Can (715–770), a well-known poet in the Tang dynasty. His poems are colorful, passionate and full of romanticism, majestic momentum and rich in imagination. The following poem is about making farewell to friend beyond the Great Wall in a snowing day. Magnificent picture of snowing like vivid blooming of pear trees impresses readers with beautiful scenery and the warmth of Spring, which make the poem unique and rich in artistic connotation and imagination.

A Song of White Snow in Farewell to
Field-Clerk Wu Going Back Home
by CEN Can

The north wind rolls the white grasses and breaks them,

Snow sweeps across the Tartar sky in August;

It's like a spring gale coming up at night,

Blooming as petals of ten thousand pear trees.

……

We see him off at the east wheel-tower gate,

Into the snow-mounds of Heaven-Peak Road

And then he disappears at the turn of the pass,

Leaving behind him only hoof-prints.

A significant poem from Wang Zhihuan (688–742, Tang Dynasty) related to Yellow River is:

At Heron Lodge
by WANG Zhihuan

The shiny sun sets by mountains,

The Yellow River converges into the ocean;

If you wish to widen your view by thousands of miles,

The best way is to go up to a higher floor.

This is a poetry by famous poet Wang Zhihuan on climbing higher for a look into a far distance, which implies that "For a grander sight, strive for further improvement." This has become a motto in China for the pursuit of the ideal state.

Li Bai has been titled as the lead among the top 8 poets during the period of Tang and Song Dynasty. He had many poems praising the Yangtze River, Yellow River, etc. The following are only typical part of them selected.

Bringing in the Wine (Extract)
by LI Bai

See how Yellow River is pouring down like from heaven,

Running forward to the ocean and never to return.

See how parents lament their gray hairs before bright mirror,

How time flies as if the black hair in morning had changed by night to white. ……

Again in this poem water is compared with time for their commonality of going forward with no return, which implies a proposal to treasure, enjoy and make good use of the present time.

A Farewell to Meng Haoran on His Way to
Guangling (Yangzhou)
by LI Bai

You have left me behind, old friend, at the Yellow Crane Terrace,

On your way to visit Yangzhou in the misty March of flowers;

Your sail, a single shadow disappears at the far end of blue sky,

Till now I see only the Yangtze River on its way to horizon.

One of China's most famous female poets Li Qingzhao (1084–1155, Song Dynasty) had her home near hot springs in Ji'nan, capital city of Shandong Province. Water topics are found in quite a few of her poems. The poem, or perhaps better the song as follows describes a fascinating scenery near the water and follows a special rhythm and a pattern originating from the Tang Dynasty, which was fully developed during the Song Dynasty:

Ru Meng Ling · Always Remember the Sunset
over the pavilion by the river
by LI Qingzhao

I always remember the sunset over the pavilion by the river,

(continued)

So tipsy, we could not find our way home.

Our interest exhausted, the evening late, we tried to turn the boat homeward.

By mistake, we entered deep within the lotus bed.

Row! Row the boat!

A flock of herons, frightened, suddenly flew skyward.

In one of her last poems, she was very melancholic. Mostly, it is due to the sudden big change of her family.

- The English and Chinese versions look like this [5]:

Qing Ping Yue • Year by Year while It Snows
LI Qingzhao

Year by year while it snows,

I often gather plum flowers,

Being intoxicated with their beauty and fondling them impudently

I got my robe wet with their lucid tears.

Having drifted to the corner of the sea and the edge of the horizon this year,

My temples have turned gray.

Judging by the gust of the evening wind,

It is unlikely I will again enjoy the plum blossoms.

And a Norwegian version of the poem (Fig. 2.4):

År for år, i snøen,
  har jeg ofte samlet plommeblomster,
  og blitt beruset av deres skjønnhet

og funnet dem uforskammende,
Fordi kjolen ble våt av deres klare tårer.
I år er jeg drevet til et hjørne
av sjøen, og til enden
av horisonten
Mine tinninger har blitt grå.
Vurdert fra kastene av kveldsvinden,
er det ikke sannsynlig at jeg igjen
vil glede med over plommeblomstringen

Finally, a modern poem is selected and has been regarded as one of the most famous poem with great momentum. It was written by Mao Zedong (also Mao Tse-tung, 1893–1976), the former Chairman of China in last century. He, besides as one of the most remarkable political leaders of the 20th century, has also been well recognized as a poet, calligrapher, and writer (Fig. 2.5).

Qin Yuan Chun: Snow (Extract)
by Mao Zedong

What magnificent northern scenery!
Thousands of miles of land sealed by ice,
Whirling snow fluttered across ten thousands of miles land.

Watching both side of the Great Wall,
Only a vast expanse of whiteness left;
The upstream and downstream of Yellow River,
Freeze its swift current.

The mountains dance like silver snakes.
The ups and downs of plateau extends like white running elephants.
Wishing to compete with the sky in height.

Till a fine day,
The land cladded in white and adorned in red
Looks more enchanting.
…

It is possibly only in the drinking songs that direct contempt is shown for water. However, we find this in a famous Norwegian student song; *D'er liddeli flaut (It is very insipid*

Fig. 2.4 The Chinese poet Qing Zhao (Li Qingzhao) lived in a small town in the Shandong Province in China–Bamai, and there we find a museum to her memory in a park with hot springs. This picture of her is located in the museum, and even the picture has a lot of water poetry

Fig. 2.5 Water lilies, is also called the nix-rose, is poetry itself and a popular subject in a lot of water poetry

to drink only water.) However, it the third verse we also there find respect for water;

> Engang holdt jeg på å krepere i Saharas ørkenland, (Once I nearly died in Sahara's desert sand)
> Jeg tok det som en mann, (I handled it as a man)
> Skjønt jeg lå på gravens rand (Even if I was on the edge of death).
> Hvor jeg snudde hue, så jeg bare himmel, sol og sand, (Where I turned my head, it was only sky, sun and sand)
>> Jeg skjønte jo sjæl (I finally understood)
>> Jeg tørsta i hjæl, (I was dying of thirst)
>> Da var det jeg ba om litt vann! (Then I asked for some WATER)
> REF: Det er liddeli flaut da gitt å drikke bare vann, (It is no good to drink only water)
>> Men blanda med Whisky kan det jo til nød gå an… (But mixed with whisky is OK)

The topic, however, is not only for student songs. Mark Twain shall have said; *Whisky is for drinking. Water is to fight for.*

More salty pleasures, and praise for water that is found in old seamen songs, in particular when they tell about the old sail ship era, is not hard to find.

The salty and simple respect to water in the first chapter Jack London's book *The cruise of the "Snark,"* is hard to find better described. We can nearly feel the water in our hair;

> Possibly, the proudest achievement of my life, my moment of highest living, occurred when I was seventeen. I was in a three-masted schooner off the coast of Japan. We were in a typhoon. All hands had been on deck most of the night. I was called from my bunk at seven in the morning to take the wheel. Not a stitch of canvas was set. We were running before it under bare poles, yet the schooner fairly tore along. The seas were all of an eighth of a mile apart, and the wind snatched the whitecaps from their summits, filling. The air so thick with driving spray that it was impossible to see more than two waves at a time. The schooner was almost unmanageable, rolling her rail under to starboard and to port, veering and yawing anywhere between south-east and south-west, and threatening, when the huge seas lifted under her quarter, to broach to. Had she broached to, she would ultimately have been reported lost with all hands and no tidings.
> I took the wheel. The sailing-master watched me for a space. He was afraid of my youth, feared that I lacked the strength and the nerve. But when he saw me successfully wrestle the schooner through several bouts, he went below to breakfast. Fore and aft, all hands were below at breakfast. Had she broached to, not one of them would ever have reached the deck. For 40 min I stood there alone at the wheel, in my grasp the wildly careering schooner and the lives of twenty-two men. Once we were pooped. I saw it coming, and, half-drowned, with tons of water crushing me, I checked the schooner's rush to broach. At the end of the hour, sweating and played out, I was relieved. But I had done it! With my own hands I had done my trick at the wheel and guided a hundred tons of wood and iron through a few million tons of wind and waves [6].

If not more salty, but at least as painting and dramatic, and a little more mythological is Alistair Maclean's gale description in *H.M.S. Ulysses*, where he in Chap. 6 *Tuesday Night*, describes what he calls *the worst storm of the war*. He writes [7];

> …At 2230, the Ulysses crossed the Arctic Circle. The monster struck.
> It struck with a feral ferocity, with an appalling savagery that smashed minds and bodies into a stunned unknowingness. Its claws were hurtling rapiers of ice that slashed across a man's

Fig. 2.6 Not "praam through the forests", but praam in the forest, which the spring is about to melt to the surface from its wintry hiding place in ice and snow, Sem Lake, Asker, Norway

face and left it welling red: its teeth were that sub-zero wind, gusting over 120 knots, that ripped and tore through the tissue paper of Arctic clothing and sunk home to the bone: its voice was the devil's orchestra, the roar of a great wind mingled with the banshee shrieking of tortured rigging, a requiem for fiends: its weight was the crushing power of the hurricane wind that pinned a man helplessly to a bulkhead, fighting for breath, or flung him off his feet to crash in some distant corner, broken-limbed and senseless. Baulked of prey in its 500-mile sweep across the frozen wastes of the Greenland ice-cap, it goaded the cruel sea into homicidal alliance and flung itself, titanic in its energy, ravenous in its howling, upon the cockleshell that was the "Ulysses". ...

But not only in the stormy seas and even in the most earthlike situations poets has found associations to the water world to express their views. Nobel literature prize winner Knut Hamsun in his book *Markens grøde* (The crop of the soil) tells about the main character Isak Sellanrå *that he walks like a pram through the forests* (Fig. 2.6).

We also include some water fascination from other Asian classic literature. Sei Shonagon was probably born about in year 963. She was a lady-in—waiting at the Court of the Japanese Emperor. Her father was a provincial bureaucrat, but best known as a poet and scholar. Sei Shonagon got a writing book from a Captain at the Royal Court. The book was used as a pillow. The notebook was filled with her observation and happenings, often very critical from her viewpoint at the Court. Not at least she is commenting on what she thinks is good or bad behavior. The book has later been named "The Pillow Book" [8].

In note number 84 she is thrilled and a bit wondering about some water observations:

> I remember a Clear Morning..
> I remember a clear morning in the Ninth Month when it had been raining all night. Despite the bright sun, dew was still dripping from the chrysanthemums in the garden. On the bamboo fences and the criss-cross hedges I saw tatters of spider webs, and where the threads were broken the raindrops hung on them like strings of white pearls. I was greatly moved and delighted.
> As it became sunnier, the dew gradually vanished from the clover and the other plants where it had lain so heavily; the ranches began to stir, then suddenly sprang up of their own accord. Later I described to people how beautiful it all was. What most impressed me was that they were not at all impressed (Fig. 2.7).

The falling snow triggers the imagination and creates association to all of us, and the poets might put the associations into beautiful verse lines with music in the rhythm of the lines, often with some melancholic undertones. Many poets have taken the inspiration and the opportunity of the falling snow as for example the Norwegian poet Arne Paasche Aasen that created the poem *Nysne faller (New snow falling)*. To translate the poem to another language would give unforgivable discredit to the original—but he fabulizes about the new snow that falls and creates a blanket on the fields and the hills. The mildness of the soft landscape, the peacefulness of the snow landscape, the white star like view both from the sky and the ground, how the snow creates the special silence, how the snow creates loneliness

Fig. 2.7 It is snowing again, and the contours soften—and quietness is sinking down over the landscape. From the Sem Lake in Asker, Norway

and the bring the mind towards the wondering of the meaning of life.

In the introduction to the short story *Det snør og snør (It snows and snows)*, the author Tarjei Vesaas gives a snowfall description about the endlessness and the hopeless feeling a heavy snowfall can give:

> It snows and snows over the plains-in an endless infinity. Not a living three. A house can be seen, only one, and it gets smaller every hour, and soon it has slowed down.
>
> Around the house the snow piles are growing. Silent and light it comes down,—the snow stars are filling the roof. No weight at all—but in the end the roof beams will crack under the heavy load.
>
> It goes on for days, where the grayness is filling the vision. It does not change; there is no end to the weather any longer. As it snows no, will it snow forever… (Fig. 2.8)

Fig. 2.8 Snow winter, Asker, Norway

Snow can also inspire to more grotesque poetry. The Norwegian author Andre Bjerke in 1958 published a collection of limericks. In translation, one of the poems sounds about like this:

> A snow landscape painter from Rody
> he painted the snow picture so the
> picture was artistically true
> and next day they do
> recover his ice-frozen body

The skill to mirror the reality, in a quiet lake or fiord can be a considerable inspiration (Figs. 2.9 and 2.10).

It looks like most rivers have given inspiration to painters and poets. Through the city of Oslo runs a small river, only a bit more than 8 km long and with a total fall of 149 m. This little river was very important for the development of the Norwegian Capital with energy to industry that later has become important companies, and it has also given inspiration to the artists that have been living in the city. Painters like Fritz Thaulow and Edvard Munch has painted from the river and quite a few poets have described the life along the river. Most inhabitants of Oslo probably know a verse from one of the songs about the river (Figs. 2.11, 2.12, 2.13 and 2.14).

Few things are as simple as water. Still, water with all its many faces and phases is sometimes complicated and difficult to grasp. Few have performed this complexity in Joni Mitchell's beautiful song better than Roger Whitaker [9];

> I've looked at clouds from both sides now.
> From up and down, and still somehow
> It's clouds illusions I recall.
> I really don't know clouds at all.

Fig. 2.9 The Ice Fiord, Møre and Romsdal County—a quiet spring day. The fiord is a perfect mirror

Fig. 2.10 Turned upside-down, the mirror picture might be just as realistic as the original

Fig. 2.11 The river through Oslo, Akerselva—at its beginning the Maridal Lake a sunny day in March. The water is now clean so there is good sight to the bottom

Fig. 2.12 In Nydalen, a former industrial area, the new buildings are architectural attempted to fit in with the old industry buildings to form a charming environment. Also the river picture has been taken care of and cleaned up, both visually and sound wise

Fig. 2.13 "Nidelven stille og vakker du er, her hvor jeg går og drømmer" (Nidelven, quiet and beautiful you are)—is the verse of a beautiful little tune. About the river that runs through the city of Trondheim in Norway. Even on a day in January, where the typical Trondheim weather is changing from blue sky to windy snow, where the snow ice comes at you nearly horizontal—it is beautiful and quiet along the river. The Cathedral—Nidarosdomen—on the other bank adds a melancholic dimension to the river view

We also bring with us typical marine expressions into the daily language, and let them express associated meanings, even if they do not necessarily have anything to do with water or the sea. Here are some examples:

Anchor	Gallion figures	Lantern
Armada	Grogg	Navigate
Ballast	Harbor	Propeller
Broadside	Virgin trip	Sail
Brigg	Caravel	Skipper
		(continued)

For full sail	Collisions course	Set course
Free harbor	Command bridge	Wave
Galeas	Convoi	

Water creates ideas, fantasy, illusions, and joy. The Norwegian newspaper VG Weekend in September 2012 [10], asked children in a kindergarten in Bergen: What is most fun with rain?

Boy 4 years: *It's fun to play with water. I use to make a river, and then I put water in it. I use sand to make the river, and dig it out, and then the water runs down the river.*

Fig. 2.14 Moscow at the end of November, 2011, photo from the window of Crown Plaza World Trade Centre Hotel before dawn while the sky was in sapphire blue, the beautiful Moscow river winds through and separates the Hotel Ukrain and Federal Government office building, which composes a harmonious picture. The moving cloud and bright vehicle lane can also be shown under long time exposure

Girl 5 years: *I'm bicycling in the rain, at full speed into the water dams. And then we are sliding better in the rain. We have a slide behind the house in the kindergarten. Have plenty of clothes on. It is fastest in rainy weather.*

Boy 5 years: *I make a volcano, with lots of sand, and there must be a hole in the top. Then I make a line downwards with my spade. I have two friends, because we really need 3 buckets, and we poor a lot of water in the volcano. The water is the lava that runs down all the sides.*

Girl 5 years: *The funniest is to have bucket and spade and water in the bucket. Then I make a coffee party. But no one can really drink it.*

Boy 4 years: *I play in a pond. There is always a water pond down the road from our house. I throw stones into it, and the splashing is fantastic.*

Water has not only inspired authors and poets, but of course also painters, photographers and other artists. We just take a short trip to Japan and to a few artists that was pioneers in their field in general, and in expressing water in particular.

The Japanese wood block print master, or rather Ukio-e artist Utagawa Hiroshige, Better known as Ando **Hiroshige** (1797–1858), was a pioneer in a number of areas regarding landscape painting, not at least with respect to creating perspective, depth, and also rain and snow in his paintings. The French painter van Gogh even tried to copy on of his rain prints—evening rain over the large bridge in Atake. Of Hiroshiges many prints, probably most well known are the series; *53 stations on Tokaido*. The print series has 55 prints from the road from Edo (Tokyo today) to the old capital til

Fig. 2.15 Print number 46 from Ando Hiroshige's Tokaido-series, Shono

Fig. 2.16 Ando Hiroshiges "Moonshine over Kazawa"

Kyoto. The 55 pictures describe stops on the road, plus the Nihonbashi—the bridge into Edo and the Saga bridge in Kyoto. In the series that was made in 1832 (some say 1833) there are 4 rain—or snow—pictures. In connection with an exhibition in Tokyo in 2000, a Tokyo newspaper claimed that these 4 prints had values 2–4 times the other print (Fig. 2.15).

Probably, the best known of the few triptych prints that Hiroshige made is *Moonshine over Kazawa (1858)*. Possibly with the exemption of the foreground, this print could have been a very nice expression of a summer night somewhere along the Norwegian coast. What we find most fascinating with the picture is the simplicity in the color design with the indigo color dominant in the fine and genius description of a summer night by the water. Symbolically, the water that dominates the picture in a proportion that is approximately the same as water dominates the globe in total (Fig. 2.16).

The great Japanese master of wood block prints/Ukioe Katsushika, also Tetsuzo **Hokusai** (1760–1849), born in Tokitaro Kawamura, is probably best known his series *63 views of Mount Fuji*. Some of the prints that has received most acclamation are "*Red Fuji*" and "*Great wave at Kanagawa*".

The wave from Kanagawa (print number 19 in the series), with its monster claws has made school for painter regarding water illustrations all over the world. The way Hokusai illustrates the underside of the wave and how the monster wave with its claws is about to swallow the two small boats with its men is unique, not only for nearly 200 years ago, but also today (Fig. 2.17, 2.18 and 2.19).

Fig. 2.17 Katsushika Hokusai's "The wave from Kanagawa" (also called "Monster wave outside the coast of Kawagawa"), with Mount Fuji in the background

Fig. 2.18 Hokusai repeated the dragon claws in another print in the series, #41, From Kajikazawa, Kai Province. The fisherman on the top of the cliff is handling his net in some impressive waves

Fig. 2.19 Hokusai's Red Fuji. Print #45 in the series—Mount Fuji in clear weather and light wind, mostly known as "Red Fuji" has little liquid water in it, but it has clouds and the snow. This is also part of a holistic glimpse into fabulous 200 years old woodblock print

Fig. 2.20 Water statue, Solar de Mateus, Duero Valley, Portugal

Fig. 2.21 This is also water, a special water as teardrop for praying for peace and commemorating the dead, sitting beside the Wall of Peace built inside the Lmjingak near DMZ, Korea with stones from 86 battlefields of 64 countries around the world

Examples about how water has been inspiration for sculptures are numerous, and are partly covered by other chapters. Here we just give an example just to remind about it (Fig. 2.20).

Water is often used for expressing human's thinking, feeling, mode, and so forth. For instance, a special water as teardrop means a lot for mankind, this is evidenced by the picture below in Seoul, Korea (Fig. 2.21).

A mythological water phenomenon that has inspired many artists is the Nix. The stories about the Nix are found in many varieties, but mainly the Nix figure is a male character that lives in water. Often we connect him with small, dark, and mysterious lakes, but the Nix figure has also been connected to brooks and rivers, larger inland lakes, and sometimes he also pops up in fiords. The Nix stories are wandering stories that has got their local colors from

Fig. 2.22 At the tourist location Huso in Hemsedal in Norway, Torbjørn Rustberggaard has made a Nix-centre to remind about the legends. *Photo* Torbjørn Rustbergard

fantastic people over time. One theory is that the Nix stories were created to scare children to keep them away from dangerous water sources.

An older dictionary tells that the Nix is a water God that is found in rivers and lakes and that can show himself as a man, a timber log or a horse. He is mean and try to fool people to the water to drown. He is particularly dangerous just after sunset.

The Nix character is found all over Northern Europe, and it is claimed that also the mermaids, the sirens and the virgins in the Rhine is a female variety of the same character.

The Nix has many names. In German—På tysk der Nix, der Neck der Nöck. In Swedish—näck, neck or stömkarl, In Danish—nøkke or åmand, In Norwegian—nøkken, nøkk, nykk, nykkj, and also fossegrim, fossekall and kvernkall.

The fantasy has been great among storytellers and artists describing the Nix.

The Norwegian painter Theodor Kittelsen has made a famous picture of a rusky head of a monster with shiny eyes in a dark little lake with rings in the water around. The picture was made around 1892 and is hanging in the Norwegian national gallery. For those who have seen this picture, they will have an imagination of the Nix for life (Fig. 2.22).

The Nix has also sometimes taken place in characters like white horses or fiddlers. The common part is that the Nix tries to fool people, and often children, to the water to drown. Most often this happens just after sunset. Persuading fiddle or harp music in quiet nights is often an ingredient in the stories.

References

1. Shouyi Bai, An outline History of China, Foreign Languages Press, Beijing, China. ISBN 7-119-02347-0, 2002.
2. http://www.zdic.net/z/1c/js/6C34.htm.
3. http://www.zdic.net/z/15/zy/51B0.htm.
4. http://www.gushiwen.org/gushi/xue.aspx.
5. Mail from Tongbo Sui 15.03.13.
6. London Jack: *På langfart med Snarken (The cruice of the "Snark"),* translated by Nordahl Grieg, Gyldendahl Norsk Forlag, Oslo, Norway, 1934.
7. MacLean Alistair: *H.M.S. Ulysses,* translated by Trond Stamsø, J.W. Cappelens Forlag AS, Oslo, Norway 1955.
8. Sei Shonagon: *The Pillow Book,* Translated by Betty Radice, The Penguin Classic, London, England, 1971.
9. Michell Joni: *From both sides now,* "New world in the morning" – Roger Whittaker, E.M.I. records, 1971.
10. Vikøyr Harald og Vågenes Hallgeir: *Fra barnemunn (from the childrens mouth),* VG helg, Oslo, Norway, 29.09.2012.

Snow and Ice 3

Over two-thirds of all the fresh water on the earth are bound in snow and ice. Ice in the form of glaciers is found on all continents, but very unevenly distributed between different parts of the world. Ice that has been created from freezing of water has a density at 0 °C of 0.916748 g/cm^3, and is thereby lighter than water. Ice will float on water, but the difference in density makes ice float with 88 % of its volume below the water surface. The volume increases when water freezes is 9 %, and this leads to expansive stresses when water freezes in a crack of rock, in pipes, in containers and in bottles. Ice that is formed by snow that has been compressed, depending on the pressure, is lighter than ice created from frozen water (Figs. 3.1 and 3.2).

Ninety nine percentage of the water on the earth that is stored as ice will be found in the polar areas. The volume is so great that it is somewhat difficult to imagine. When we compare the ice area in the Antarctic of about 12 million km^2 with the area of the oceans in the world of 86 million km^2, we see that for each 7–8 m of ice that melts in Antarctic, theoretically the oceans level will rise with nearly 1 m. The very theoretical estimation of what the ocean level will rise with, if all the ice in Antarctic melts, shows a result of about 50 m. In comparison, calculations have been made on the effect of melting of all the ice in Greenland showing that the oceans will rise with 6 m.

About 10 % of the ice in Antarctic is floating. The glaciers occasionally are calving large mountains of ice that can be up to 300–500 m deep. The sea ice in Antarctic is partly fixed to the bottom. If the ice covered area in west Antarctic looses from the sea bottom, it will increase the sea level by 16–50 cm. The glaciers then will float and will increase the sea level by 6–7 m [2].

It is hardly difficult to understand and accept that Antarctic is the coldest and most windy area on earth. More special is the fact that even with all the ice (water), Antarctic is the driest continent with the least amount of precipitation. The continent is the fifth largest, covering 10 % of the area on the earth, whereas 98 % is covered by ice glaciers, in some places up to the height of 4 700 m.

Seven nations have claimed parts of Antarctic; Argentina, Australia, Chile, France, New Zealand, Norway, and Great Britain, and the continent is divided up in a sector principle.

Protection of the environment south of 60° S is secured through the Antarctic Treaty that came into force in 1961. Twelve countries signed the original treaty. Later more countries have been added. In 1991, 40 nations signed "The Environment Protocol for Antarctic." According to this protocol, all searching for and excavation of minerals in Antarctic are prohibited for 50 years [3].

The ice on Greenland, the sea ice in the Arctic Sea and a few continental glaciers are originating on the northern hemisphere from the ice age some 8 000 years ago, and these have now started to melt in at a disturbing speed. The ice on Greenland has enough volume to increase the sea level by 7 m. In the summer of 2002, this ice massive was reduced by one million square kilometers, and in 2004 it was discovered that the ice was melting 10 times faster than originally expected [4]. In an interview with the Norwegian newspaper VG on January 6, 2013, Director Jan-Gunnar Winter at the Norwegian Polar Institute 2013 [5], tells that in the summer of 2012 it was recorded with melting on 97 % of the Greenland Ice, and even at a height of 3000 m above sea level. Last time something like this had happened was in 1889.

The most imposing change of the glaciers happens in the Mount Everest area,—the Roof of the Earth. China Central Television (CCTV) launched a series of TV programs titled as Story Board: Take a Deep Breath—showcasing the great challenge of China by listing lots of alerting figures and facts on various subjects, for instance; the global warming, the depletion of water, the air pollution, etc. A few data below are extracted from a two-episode English program subtitled as Waiting for winter on Mount Everest [6];

- Glaciers on the Qinghai–Tibetan Plateau in the past 30 years, every year 131.4 km^2 disappeared.
- In recent 30 years, every year the snowline retreats 150 m.

Fig. 3.1 The local Christmas tree needs snow to create the right mood

The interview of local residents at Rongbuk Monastery also proves that in recent decades the annual temperature difference has narrowed each year. Winter in particular has become warmer and sometimes it is hard for them to preserve meat in the winter as usual. More rocky surfaces are exposed by the dwindling glaciers and melting snow. The existing ice used to be very dense and shiny, but now becomes loose and fragile. These are attributed to the global warming in world context, and to the other the increasingly more mankind activities including mountaineers in a regional view.

No wonder the local old's saying goes, "Never walk over Everest as you will cause the holy mountain to shrink."

Scientists predict that all the glaciers on the Qinghai–Tibetan Plateau will have melted in the next 30 years.

Fig. 3.2 Approximate area of the glaciers in world according to the USA's National Snow and Ice Data Center [1]

Area	km²
Antarctic	11 965 000
Greenland	1 784 000
Canada	200 000
Central - Asia	109 000
Russia	82 000
USA	75 000 (incl. Alaska)
China	33 000
South - America	25 000
Iceland	11 260
Scandinavia	2 909
The Alps (Middle Europe)	2 900
New Zealand	1 159
Mexico	11
Indonesia	7,5
Africa	10

The statistics of the Meteorological Bureau of Qinghai Province indicates that since 1980s temperature rise rate in Tibetan Plateau for every 10 years is 0.35 °C—far higher than the global temperature rise rate per 100 years of 0.74 °C.

Experts predict that in the next 7 years, in the Qinghai–Tibet Plateau, the permafrost thickness of less than 10 m will disappear.

The sea ice in the Arctic Sea also melts quickly, and in September 2012 the first commercials sailings through the Northeast Passage took place. In 2012, 48 ships sailed through the passage. There were 44 more ships than in 2010 [5]. In the interview with Winter in VG mentioned earlier, he says that the most pessimistic climate models now claim that we will have an ice-free polar sea already within 10 years, while the more moderate models anticipate ice-free waters sometime between 2050 and 2070. The probability for increased sailing through the Northeast Passage from Asia to Europe in large parts of the year increases. This positive possibility has its price in others and negative climate consequences.

Measuring with submarines has shown that the ice thickness of the sea ice is only 60 % of what it was 40 years ago.

The Northeast Passage was sailed through for the first time in the years 1878–1879 by Swedish—Finnish Adolf Erik Nordenskiold with his ship "Vega." Then it took 40 years before the next through sailing, when Roald Amundsen in 1918–1920 sailed through with the polar ship "Maud." "Maud," however, was built in the small idyllic village Vollen in the community of Asker in the Oslo fiord. The boat building tradition at Vollen goes far back in history. The place is also well known for building of many famous competition sailors (Fig. 3.3).

Roald Amundsen, from Norway, with a crew on board the polar ship "Gjøa" was in 1906, after 3 years of hard work, the first that sailed north of Canada through the Northwest Passage. "Gjøa" was built in Rosendal on the west coast of Norway in 1872. After the Northwest Passage journey they sailed "Gjøa" on to San Francisco where the ship was on exhibition in the Golden Gate Park in many years. In 1972 the boat was brought back to Norway (Fig. 3.4).

In 1910, two Norwegians, Børge Ousland and Thorleif Thorleifsson sailed through the Northeast Passage and the Northwest Passage in one summer in a 31 foots glass fiber—trimaran. A fantastic journey was very well described in their book *Seilas rundt Nordpolen (sailing around the North Pole)*, again another frightening proof of what is about to happen regarding the ice melting in the polar areas.

The ice melting in the northern regions has considerable interest for the shipping industry. In connection with a shipping conference in Oslo in June 2013, the newspaper VG [5] tells that with an ice free Northeast Passage, ships

Fig. 3.3 Idyllic Vollen, a sunny day in March

Fig. 3.4 Autumn 2012—A house is built around "Gjøa" at Bygdøy in Oslo

can save 13 days on a trip to Japan, 11 days from Kirkenes in Norway to South Korea, and 11 days from Kirkenes to China, compared to sailing through the Suez Canal. The article claims that the northern route in some cases can give a time saving of 40 % and 20 % of fuel savings.

The main source for glaciers is snow. Over time the snow is pressed together, and together with frozen melting water it transforms to ice. This ice is lighter and less homogeneous than the ice that freezes on lakes in the winter.

The age of the ice might vary from a few years to several hundred thousand years. This fact or property has made the glaciers very interesting for researchers, as they can drill out cores, and in studying these cores, they can make estimates of the climate far back in time.

Norwegian glaciers represent a considerable part of the Scandinavian glaciers. However, 18 of the largest glaciers in Norway, is located at Spitsbergen. Here we also find the largest glacier in Europe—Austfonna.

In total, mainland Norway has some 1600 glaciers, whereas 900 in Northern Norway. Sixty percentage of the glazier areas is however, in southern Norway. In total the glaciers covers about 1 % of the Norwegian land area [7].

The largest glacier in Norway, Jostedalsbreen, is also the largest in continental Europe. The height at the top varies from 1200 to 1500 m above sea level. Each year, the glacier receives about 3–6 m of new snow.

Jostedalsbreen is the source for the water ways in Loen, Olden, and Stryn, and also gives power to the hydropower station in the Jostedal waterway (Figs. 3.5, 3.6 and 3.7).

The Alps in middle Europe are the other main glacier areas in Europe, after Scandinavia, and Switzerland is a central country in this respect. Typical for the largest glaciers in Switzerland, is that in addition to being closer to equator is that they are on a much higher altitude than the Scandinavian glaciers (Fig. 3.8).

Aletsch is the largest glacier with an area representing 13 % of the total Swiss glacier area. In addition to the 4 largest glaciers, Switzerland has [8]:

- 6 glaciers with areas between 20 and 30 km^2
- 10 glacier with area between 10 and 20 km^2
- 24 glaciers with areas between 2 and 10 km^2.

In an article in December 2008, Jonathan Amos [9], tells that the Swiss glaciers melt at a disturbing speed. With reference to a stately Swiss research institute he claims that in 2008, they had recorded that 10 km^3 of ice was lost from

Fig. 3.5 It is not only as a water magazine we utilize the ice. It is in the morning after Christmas Eve—possibly it is the Christmas presents that is in use for the first time. The clearing truck has already cleared the snow from last night

Fig. 3.6 Snowfall, rain and frost in a few days give a dangerous surface on the walkways

the all together 1500 Swiss glaciers over the last 9 years. In total, the Swiss glacier volume in 1999 was 74 km³. Recordings on a representative number of 30 glaciers showed loss in glacier thickness of 1 m per year. The main reason is that the melting season is longer than the frost and snowfall season. Worst case was under the heat wave in southern Europe in 2003, when 3 % –4 % of the volume of the Swiss glaciers were lost.

As mentioned, the glaciers are the world's by far largest fresh water reservoir. In some areas of the world, the glaciers are also direct water reservoirs for many people.

A number of the most important rivers in Asia have their beginning in the Himalayas, the highest mountain range in the world with 9449 glaciers as the world largest glacial area outside the Arctic. Where melting water from the glaciers is of uttermost importance to the water supply for billions of people, not at least in heavily populated countries like; China, India, Pakistan, and Bangladesh (Fig. 3.9).

Alarming reports tell that the glaciers in Tibet is melting with a speed indicating that the glaciers will half their size in the next half of hundred years (Fig. 3.10).

The Himalaya glazers are important water sources for the river (Fig. 3.11):

- Huanghe (The Yellow River)
- Changjiang (The Yangtze River)
- Brahmaputra
- Ganges
- Indus
- Irrawaddy
- Mekong
- Salween (Thanlween)
- And a number of smaller rivers

These rivers have an important influence on water supply, electricity production, transport systems, living conditions and economy in the countries they run through (ranged after their number on the world population statistics) [10]:

1.	China	1 347 million
2.	India	1 210 million
6.	Pakistan	810 million
8.	Bangladesh	153 million
13.	Vietnam	88 million
20.	Thailand	65 million
26.	Myanmar	49 million
45.	Nepal	27 million
69.	Cambodia	14 million
104.	Laos	6 million
164.	Bhutan	0.7 million

NAME	AREA km²	COUNTY
Jostedalsbreen	487	Sognog Fjordane
Vestre Svartisen	221	Nordland
Søndre Folgefonna	168	Hordaland
Østre Svartisen	148	Nordland
Blåmannsisen	87	Nordland
Hardangerjøkulen	73	Hordaland
Myklebustbreen	57	Sognog Fjordane
Okstindbreen	46	Nordland
Øksfjordjølelen	41	Finnmark, Troms
Hardbardsbreen	36	Sognog Fjordane
Salajekna	33	Nordland
Frostisen	25	Nordland
Gihtsejiegna	25	Nordland
Sekkebreen / Sikilbreen	24	Oppland
Tindefjellbreen	22	Sognog Fjordane
Simlebreen	21	Nordland
Tystigbreen	21	Sognog Fjordane
Holåbreen	20	Oppland
Grovabreen	20	Sognog Fjordane
Ålfotbreen	17	Sognog Fjordane
Fresvikbreen	15	Sognog Fjordane
Seilandjøkulen	14	Finnmark
Smørstadbreen	14	Oppland
Strupbreen / Koppangbreen	13	Troms
Gjegnalundbreen	13	Sognog Fjordane
Veste Memurubreen / Hellstugubreen	12	Oppland
Spidstinden	12	Nordland
Storsteinfjellbreen	12	Nordland
Jostefonn	12	Sognog Fjordane
Midtre Folgefonna	11	Hordaland
Veobreen	9	Oppland
Langfjordjøkulen	8	Finnmark

Fig. 3.7 Norwegian glaciers [7]

Fig. 3.8 The largest glaciers in Switzerland [8]

Glazier name	km²	Length 1850	Length 2010	Highest point	
Aletsch	117,6	26,5 km	23,6 km	4 193	Aletschhorn
Gorner	63,7	16,0 km	14,5 km	4 634	Monte Rosa
Flesh	39,0	17,1 km	14,7 km	4 273	Finsteraarhorn
Unteraar	35,5	14,6 km	13,9 km	4274	Finsteraarhorn

Fig. 3.9 From Tibet in June

Fig. 3.10 Brahmaputra (Brahma's son) has its origin from the glazier Jima Yangzong Chemayungdung at a height above sea level of 5590 m in south western Tibet, China. The river then runs from west towards east in about a height of 3 600 m over the high planes in Tibet, not far from Lhasa (The picture). It then floats through India and Bangladesh and runs into the sea in the Bengal Bay after 2 900 km. The width of the river is from 3 to 18 km

The changes in the glazier melting are expected to bring with it increased flood activity as well as increased drying periods, and obviously changes that will be reflected in the economy and development for these countries representing more than half the population on the earth.

The illustrations below show what is going to happen with the mega river Indus in terms of increased water, and/or reduced water over time, depending on what changes can be expected in changes in middle temperature. The irrigation systems, depending on water from the Indus River, are claimed to the biggest in the world, and changes in the water volume can have catastrophic consequences for one of the most populated areas in the world, and have drastic consequences for the economy of Pakistan. Increased melting of

Fig. 3.11 The Himalaya Rivers

the glaciers first leads to increased flooding, but as the glaciers get smaller, to lower water volumes (Fig. 3.12).

In a number of areas in the world, the seasonal large supply of snow will be an advantage, but this also creates challenges. The snow represents important supply to brooks, rivers, and power stations, while the snowfalls can also represent challenges for the road and railway traffic (Figs. 3.13, 3.14 and 3.15).

Snow has insulating properties. One effect of this is that there is less frost in the ground, and some places do not exists at all if there is enough snow. However, where the snow has been cleared on the roads, the frost in winter can easily in a country like Norway go 1–2 m down in the ground. The isolating effect of the snow also has major effects for the animal and plant life that exist under the snow. When the spring comes we might observe an interesting phenomena along the snow cleared roads. The snow pile cutting might be nearly vertical on the south side of the road where the sun has little effect. On the north side, however, the inclination gets smaller and smaller as the melting progresses. At the same time, the road surface gets wider and wider on the north side.

Frozen ground or water in the ground that has frozen to ice vary in thickness with various conditions—the ability of the soil to keep the water or to drain it away, the isolating properties of the soil and the possible snow above, the amount of cold, and various heating elements. The depth of frozen ground therefore varies from year to year and very much geographically. The frozen ground gives us a number of challenges—foundations for buildings and other structures have to be isolated and made deep enough, ground heave in roads that does not have proper foundations, more challenging work in the winter, etc. (Fig. 3.16).

When the first snow comes in a typical "winter country," most people have been prepared for it for a long time (Figs. 3.17, 3.18, 3.19, 3.20 and 3.21).

A chapter about snow and ice cannot be written without mentioning Norway's national sport—skiing. According to the Skiing museum at Holmenkollen ski stadium in Oslo, skiing can be documented 4000 years back in time in Norway, Sweden, Finland, and Russia. In particular in moors, skies, or reminiscences of skies have been found a number of places. The oldest reminiscences of skies have been found at Drevsjø in Nordland County, Norway and they are 5 200 years old (Fig. 3.22).

We know that the Vikings some 1000 years ago had high appreciations for skiing abilities, and that skiing had its own God—Ull. Ull's skiing abilities could only be matched by the skiing Goddess—Skade. Then it is probably not so strange that in the modern skiing world that the modern Norwegian Skiing Goddesses are even more popular that the modern male skiing Gods. Norway are building out "skiing temples" for various types of skiing on the white water over

Fig. 3.12 Changes in the water volume over time in the Indus River, depending on annual temperature changes. *Source* UNDP-report 2006 [11], reference to World Bank 2005

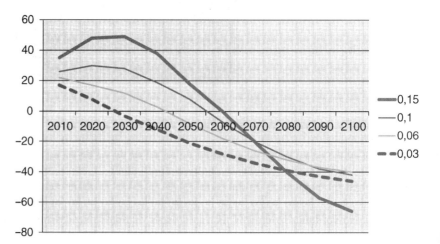

Fig. 3.13 The frozen water in a reservoir keeps its original form of waving and shiny surface. January 2009, North suburb of Beijing, China

the whole country, and in 1994 an Olympic temple was built in Lillehammer for entertainment and worship for a greater part of the world.

But, what is called "the Mekka of skiing" is Holmenkollen just outside Oslo. This is the most holy of all skiing temples in the world, and has its main competition in the middle of March each year. People go there the rest of the year too of course, in particular tourists that want to experience what is more Norwegian than anything else (Fig. 3.23).

The Norwegian neighbors to the east, the Swedes are smiling about the Norwegian affection for skiing and their imagination of being born with skies on their feet.

Skiing experience is different for most people, man or probably most Norwegians think of skiing first of all as a recreation trip on cross country skies up in the mountain or a sunny winter day in the nearby forest area (Figs. 3.24, 3.25 and 3.26).

Ice Industry and Ice Export

If we mention ice industry today, most people will probably think about ice cream. This was not the case a decennium ago.

The 4th of February 1900, the Chairman of Asker Ice workers Association sent out a message to 20 000 fellow workers in Norway about founding the Norwegian Ice workers union. It was sent to a group that he called "one of Norway's largest industrial branches". In USA it was anticipated that there were about 90 000 persons working in the ice industry towards the end of the 1800s [12].

The history behind "harvesting" of natural for the purpose of refreshing food and drinks has somewhat different importance and focus in different sources, but it seems that there are reasonable agreement about that the oldest examples can be localized to China about 500–1000 years BC, or just before the first Chinese dynasty that represents a complete China.

The earliest record in China on using ice can be dated back to Xia Dynasty (ca. 2070–1600 BC). Already at that time, ice was so precious in summer that it was used as the emperor's award to the minsters, and specific officials were assigned to take care of preparing and preserving ice. Ice was cut in the winter and stored in ice houses or underground cellars, often covered and isolated by sawdust—for the use in the warmer periods of the year [12]. It is also said that a kind of ice cream-like food was invented in China

Fig. 3.14 Snow in Beijing has been getting rarer. Can you imagine in the later time of the year for 108 days from October 2010 through February 2011 Beijing had suffered from a period of no precipitation? The average temperature of Beijing in February was minus 0.8 °C, which is 0.6 °C higher than usual. *Picture* January 2010 Summer Palace, Beijing, China

Fig. 3.15 Newly opened road after the winter season, Sogn Mountains, Norway

Fig. 3.16 Digging ditches in the winter. The machine must cut through the hard frozen top layer. Here they have made a small fire with coal to melt the frost and ease the digging operation

about 200 BC with a milk and rice mixture frozen by packing it into snow. And in Song Dynasty (960–1279), ice-cooled food and beverages had already been very popular. The largest scale of preserving ice in Chinese history recorded in Qing Dynasty (1636–1912), which in Beijing at least 14 big ice cellars had been built. Just in the Forbidden City only there were 5 big ice cellars storing 25 000 pieces of ice blocks mostly for the daily use of the emperor, which are equivalent to about 60 thousand cubic meters of storing capacity, based on author's calculation (Fig. 3.27).

Wikipedia also mentions that the Persians about 400 BC, knew the art of storing ice in the deserts in the middle of the summer, that they collected the ice from the nearest mountains in the winter. They also invented a special chilled food, made of rose water and vermicelli, which was served to royalty during summers. The ice was mixed with saffron, fruits, and various other flavors [14].

Reports about examples from India, Egypt, various Mediterranean countries, North Africa, Russia and from South America are also found before the big ice export adventure started in USA in the beginning of the 1800s.

Before 1830, the "living time" for food was usually prolonged through salting, smoking, canning, and similar conservation methods. The butchers normally use only slaughtered animals for sale in the same day, and milk was transported to the dairies mostly during night time when the temperature was lower. Wastes and losses from fruits and vegetables were considerable. Large shopping from fresh food was nearly impossible for ordinary families. The trade with ice in a larger scale is claimed to have started with the businessman Frederick Tudor in New England in the USA in 1807 [12]. He started up shipping ice to rich Europeans on islands in the Caribbean. The ice trade really picked up speed in the 1830s when Todur, in addition to supplying ice

Fig. 3.17 The first snow is falling, and the special shops are prepared with the clearing devices ready for sale

Fig. 3.18 When the snow falls, there might be some constraints in our daily life

in the USA, also shipped ice to England, India, South America, China, and Australia. Already in the middle of the 1800s, ice was normal for the people living in New York.

Between 1840 and 1850 this had grown to a global industry. USA was the market leader, while Norway in a global context was "little brother," but along the coasts in Europe and in England in particular was the dominant country. Wallin Weihe and Syvertsen [15], even claim that Norway had a market share of 99 % in these markets. The first ice export from Norway to England took place in 1822, with ice in barrels from the glacier Folgefonna i Sørfjorden in Hardanger on the west coast. The ice industry and the ice export, however, did not increase considerably before about 1865, and then ice saws had also come into use. The top years for the ice industry, ice trade, and ice export were in the years from 1890 to 1900. Then the activity reduced drastically in pace with the development in artificial ice production and the development of refrigerators. Nearly full stop in the export came during the Second World War. From about 1915, the export was down to below 1/20 of what it was in the top years. However, it took a much longer time before the use of natural ice completely disappeared. As late as in the 1940s and a bit into the 1950s, the milk shops in Oslo got weekly delivery of ice blocks to their ice boxes. Some activities in Norway even used natural ice as late as in the 1970s.

The top year for Norwegian ice export was probably from 1888 to 1900, with export of more than 1 million tons of ice

Ice Industry and Ice Export

Fig. 3.19 The white snow changes the character of the landscape and is damping the sound. The mood at a churchyard at winter time is special

Fig. 3.21 Snow shelter, Japan. On the northwest side of the Honshu Island in Japan, the snow falls are considerable in the winter. When the warm, humid winds from the Pacific Ocean meet the winds from Siberia on the west side of the mountains, experiences with 5 m of snow get normal. Snow shelters on roads and railway lines become an obvious tool to improve against the moods of the nature

Fig. 3.20 Not only has the landscape changed character when the white water arrives. Structures and figures are giving new associations

per year. The year of 1884 also seemed to have been a good export year. An important reason for large variations from one year to the other was the weather conditions in England and on the European continent. Cold winters and summers lowered the import needs and the prices. The price variation from one year to the other was formidable. Price differences from 10 to 1 have been recorded from delivery of good winter ice in the summer on a top delivery year to spring ice delivery with a lot of cold in England and Germany. In the

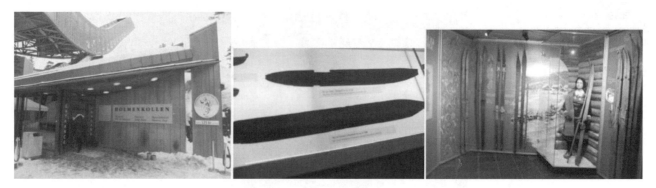

Fig. 3.22 The skiing museum at Holmenkollen shows skies from Hedemark and Finmark Counties from year 600 to 1200, and skiing traditions from Telemark County

Fig. 3.23 View from the tower of the Holmenkollen ski jumping hill a Saturday in January is a tourist attraction. The Oslo Ski festival is going on with cross-country competitions for children and juniors. This skiing stadium is in activity during the traditional Holmenkollen competitions in March

years with the lowest prices there were probably not much profit for the export companies. The largest ice company in England was Lefwich Co., and they had ice storage houses that stored 900 000 kg of ice at any time to be able to meet the market variations [14]. The most important export country without doubt was England, but it was also exported ice from Norway to Germany, West Sweden, Denmark, Netherland, Belgium, France, and to a number of countries around the Mediterranean Ocean. Exotic stories even tell how the sail ships from Norway were transporting ice as far as Bali in Indonesia so that the rich people could have ice in their drinks before supper. Recordings are found about export of ice from Risør about 200 km south of Oslo to Paris as early as in 1825 [16]. Germany was for many years a good number two as import countries of ice from Norway.

It was exported ice from a greater part of the Norwegian coast,—the Trondheim fiord and southward, that made the southern coast and eastern part of Norway, as the industry developed, became the most important export areas. Closeness to the coast, shipping traditions, shorter sailing way to the markets, and the possibility for return freight were important reasons for this. My father told about a period as a sailor around 1908–1909, when the voyage went with ice from Asker near Oslo to Liverpool in England and with coal as the return freight.

As the export developed, many dams, ponds, and small lakes were built artificially with short distance to loading of ships. A number of these dams or ponds exist even today. They are often located in flat areas with possibilities for a sloping transport channels down to the sea. Typical depths are about 2 m. The ice blocks that were cut were heavy. A standard block was $2 \times 2 \times 2$ feet, but they could be higher when the ice was thicker. The saws that were used were about 2 m long with 5–8 cm long teeth (Fig. 3.28).

The area around the towns of Brevik and Kragerø seems to be the most active area for ice export. For Kragerø, the ice

Fig. 3.24 Tempting white cross-country skiing track at Myggheim in the Asker forests, Norway, a sunny day in March

Fig. 3.25 Norwegian champion Petter Northug in snowy weather fighting Dario Cologna from Switzerland and others during the Holmenkollen 50 km in 2012, the last 50 km race in World cup circus of the season

export was of particular importance, as it was considerable reduction in the lumber export, and this was also in a difficult transition period from sailing ships to steam ships. In periods the value of the ice export was up to one-third of the timber export from Norway [16]. In the three communities Solum, Eidanger, and Bamble, nearly 1300 men were working for 20–25 weeks in the winter 1895–1896. The sail ships, however, were used in the ice trade for a long time. Kragerø was a center for the ice trade, we can find proof there even today as we hardly find so many English plants as in any other places in Norway. The reason can be traced back to ice-ship ballast that had to be dumped when they were taking in another ice load [17].

Also in the inner Oslo fiord and in the County of Akershus the ice industry was important. In the community of Asker is recorded 22 old ice ponds, and in the community Nesodden 25 ponds. Many of these ponds, dams, or lakes were artificially established for the course. Toward the end

Fig. 3.26 15. March in front of Asker Town Hall. The winter goes towards its end and the Holmenkollen skiing competition is over, but it is snowing again, covering the old dirty snow piles with a new white color

Fig. 3.27 Illustration of ice block making in Ancient China [13]

of the 1800s, 94 000 tons of ice were exported from Akershus, and 53 000 tons of ice from Asker alone.

The use of ice by ordinary consumers also increased, and the national consumption was an important reason for the ice industry to continue when the export went towards its end. The poor country, Norway, was far slower in social modernization and use of new and modern freezing technology and refrigerators than countries like England and Germany.

Ida Vesseltun tells that "Christiania Is-Magasin med fabrikk for Isbeholdere (Christiania Ice—Magazine with factory for Ice containers" (Christiania is an older name for the city of Oslo) was established in 1857, and they claim in their catalogue from 1885 that they were the first in Norway in this business [17]. In their catalogue they tell that there in 1859 were 3 "Iskister" (ice boxes) in use in Christiania, and that this grew to 2100 in 1884.

In a number of occasions it has been mentioned that it was the upper class and the rich that could afford ice in the summer in England and Norway. There might be a lot of truth in this, but it is not quite correct.

At *Christiania Portland Cementfabrikk* at Slemmestad near Oslo, the working regime for the workers in the office was fairly strict about 100 years ago, measured after today's standard, all the workers in office were provided with an icebox, and they got one ice block delivered home every week. Some of the explanation might be found in the fact that the factory had their own ice business. The factory was established in 1888, and the production started for real in 1892, but there were no profit in the first year. The first year they recorded a reasonable profit was in 1898. That was the year that they started with ice export from the Torp dam or lake [18].

Fig. 3.28 Ice saw and ice clip pictured outside Røyken History Associations house "Kornmagasinet" in Røyken, Norway. In this house the local history association has collected a number of tools, etc., from the local ice industry

It was not only a small "side-business" that Christiania Portland Cementfabrikk had in the ice activity. Ida Vesseltun [17], tells that the factory took out ice from the Torp dam. During the third and second last week of the ice activity in the winter of 1898 they had 53 men and 31 horses working.

Fig. 3.30 Modern people want to have ice in the house, whether it is natural ice, or it comes from the refrigerator. Ice is a water alternative that more and more people take for granted

Of these 10 were working with cutting, 19 horse transport, 8 with storing and the rest with other kinds of work;—foreman, road repair, other transport, snow clearing etc. The normal day had 10 working hours. In addition to this crew, came the youngsters that were hired in during winter for snow clearing (Fig. 3.29).

The ice trade in the world has set its footprints in many ways. The worldwide well-known shop concept 7—Eleven originates from the Texas-based company Joe C. Thomson Southerland Ice Company [16].

The natural ice industry has been taken over by ice machines, freezers, and refrigerators of various formats and seizes. An Internet search in April 2013 gave more than 77 million hits in 0.22 s on the search word *Ice machine.*

Fig. 3.29 Høymyrdammen lake at Slemmestad, was a water source for the cement factory at Slemmestad for more than 50 years. This was also one of the 22 ponds in the district. In the background a housing area established in later years

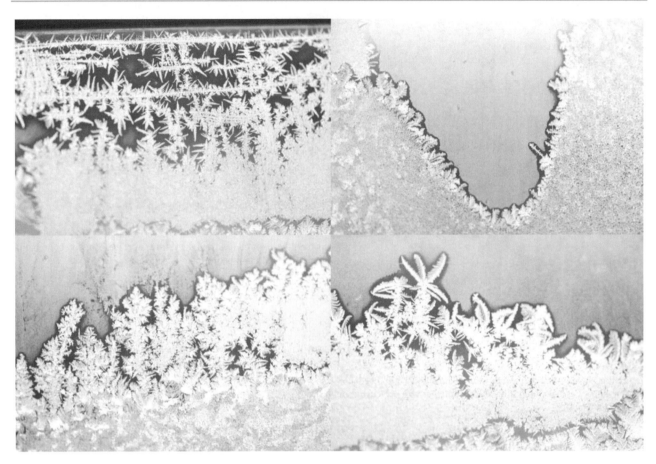

Fig. 3.31 Naturally growing ice plants/forest in the cold window glass in winter in Feb. 2010 in a coastal town Rushan in Weihai, China during the Spring Festival, which are similar as Fig. 1.19 in Chap. 1, but quite different in the patterns of ice plants or forest indicating the fascinating morphology ice crystal during phase change of water. This certainly adds the celebration during the Spring Festival, the most important Chinese New Year

The idea about cooling and freezing water is not new. The first solutions were shown in Glasgow, Scotland as early as 1748, and the first commercial refrigerator for household use was presented in 1913. However, it took more than 50 years from this until refrigerators came into common use for everyone in countries like Norway.

The way Vietnamese drink hot tea with straw in summer is also very interesting. This seems difficult, yet if you put some ice cubes in your tea cup, the hot water is cooled down to mild temperature when it is sucked and passed through the floating ice layer. Now you can have the "Ice" tea with straw. Would you like to try this?

Alternative ways to use natural ice haves created a lot of ideas. Residues from floating ice bergs from Greenland have been harvested on the east coast of America and sold as bottled for good prices, and projects ideas with the aim to improve the ice and water situation in Saudi Arabia by towing ice bergs from the Antarctic are no longer new (Fig. 3.30).

Finally let us relax and entertain ourselves with some interesting pictures related to ice and snow (Figs. 3.31 and 3.32).

Fig. 3.32 Cat came out to step on the snow and enjoyed the scenery February 2009, Beijing Botanical Garden

References

1. Internet 30.09.12: USA – National Snow and Ice Data Center: *All about glaciers,* http://nsidc.org7cryosphere/glaciers/questions/located.html.
2. World Watch Institute: *State of the World 2012.* Island press, Washington DC, USA 2012.
3. Hågvar Sigmund og Berntsen Bredo: *Norsk Naturarv (Norwegian Nature Heritage),* Andresen & Butenschøn AS, Oslo, Norway, 2001.
4. Henriksen Thor Harald og Mortensen Terje: *Ekstrem Isrekord (Extreme Ice Record),* VG, Oslo, Norway, 9. januar 2013.
5. Bjørn Haugan og Anders Bergan: *Sparer 13 dager til Japan (Saves 13 days on the way to Japan).-* VG, Oslo, Norway, 5. June 2013.
6. January 27–28, CCTV News, Story Borad: Take a Deep Breath - Waiting for winter on Mount Everest. http://search.cctv.com/search.php?qtext=Take+a+deep+breath+waiting+for+winter+of+mount+everst&type=video&sid=0000&pid=0000.
7. Internet 30.09.12: Wikipedia: *list of glaciers in Norway.* http://en.wikipedia.org/wiki/list_of_glaciers_in_Norway.
8. Internet 20.03.13: *List of glaciers in Switzerland.* http://en.wikipedia.org/wiki/List_of_glaciers_in_Switzerland.
9. Jonathan Amos: *Swiss glaciers in "in full retreat",* Internet 20.03.13. http://news.bbc.co.uk/2/hi7science/nature/7770472.stm.
10. Internet 08.09.12: Wikipedia: *List of countries by population.* http://en.wikipedia.org/wiki/list_of_countries_by_population.
11. UNDP: *Human Development Report 2006, Beyond scarcity: Power, poverty and global water crisis,* United Nations Development Programme, New York, 2006.
12. Internet 12.04.13: *Ice History.*
13. *Yangcheng Evening News,* Sept. 3, 2010, http://news.xinhuanet.com/shuhua/2010-09/03/c_12515045.htm.
14. Internet 12.04.13: *Ice trade,* http://en.wikipedia.org/wiki/Ice_trade.
15. Hans-Jørgen Wallin Weihe and Carsten Syvertsen: *Den norske iseksporten (The Norwegian Ice export),* Hertvig Akademisk, Stavanger, Norway, 2012.
16. Internet 14.04.13: *Isdrift,* http://no.wikipedia.org/wiki/Isdrift.
17. Ida Vesseltun: *Det er verre for han som holder i den andre enden av saga,* Hovedfagsoppgave, Avdeling for etnologi, IKS, Universitetet i Oslo, Norway, 1994.
18. Frithjof Gartmann: *Sement i Norge – 100 år(Cement in Norway – 100 years),* Norcem AS, Oslo, Norway, 1990.

The Oceans

The word OCEAN is connected with mystery, respect, and infinity in it. We imagine something big, and endless, a bit further than human imagination. The forces connected to it are imagined as enormous, and in reality that are greater than what we have of comparable experience.

A lot of Chinese idioms since ancient time have been used to characterize the vastness or infinity of sea. For example, idiom 福如东海[fú rú dōng hǎi] means Happiness as immense as the eastern sea. Idiom 情深似海 [qíng shēn sì hǎi] describes that the deep feelings, love, or affection are so deep like the sea that cannot be measured. The earliest and probably the most famous poem about the sea in Chinese history was written by Cao Cao (155–220), a famous statesman, military strategist and writer, and calligrapher of the Eastern Han Dynasty (25–220). He was also the main founder of the Wei regime of the Three Kingdoms, first as the general and then prime minister of the Eastern Han Dynasty. He later become the king of Wei regime. He made great contributions to the unified central plains of China.

Here is his **The Sea View**.

The Sea View by Cao Cao
Climbing Jieshi Mt. to the east to enjoy the sea
How rippling the water waves, with mountain and island towering
Forested by trees and a variety of flourishing grass
The autumn wind rustles in the air, the sea churns up giant waves
The sun and the moon go up and down like from the sea
The starlight of Milky Way seems to be produced from the sea
What a wonderful life! Here is the song to express my ambition

But, sometimes we can experience the ocean with a sun-mirrored idyllic surface full of illusions and exciting experiences and dreams, possibly of other and even more exotic beaches (Figs. 4.1 and 4.2).

Underneath this shining, slightly movable, and a bit rugged surface that we experience with the ocean, there are mountains and valleys, plains, rock formations, and reefs that we associate with landscape rather far from the ocean front. Some peaks are sticking up over the surface, and we call them islands, and possibly reefs if they are low. In more than 600 m depth, the water is colder than the freezing point on the surface. The largest ocean depth sticks down about what is comparable to putting the Himalayas and the European Alps on top of each other. The largest depth is the Marianas-ditch in the Pacific Ocean, about midway between Japan and New Guinea. The ditch is about 2 550 km long (approximate as the length of the country of Norway), and has an average width of 69 km. The maximum recorded depth is 10 994 m [1]. Some unconfirmed recordings from the southern end of the ditch claim depth of 11 033 m, The Challenger Depth [2] (Figs. 4.3 and 4.4).

What type of life that exists on these large depths is still a bit unclear. The pressure is formidable, 1000 times the pressure we have on the surface of the earth, and a normal submarine would be pressed together in an implosion. James Piccard and Don Walsh reached the bottom of the Marine ditch in 1960 with the special submarine *Trieste*, and claimed to have seen both swimming shrimp-like animals and flat fish on the bottom. In March 2012, the film producer James Cameron (made the film *Titanic* in 1997) repeated this fantastic deep diving with the special submarine *Deepsea Challenger*. They also observed shrimp-like animals and amphipods and single celled animals. Researchers from amongst others Denmark claim that there might be special pressure resistant bacteria on the bottom that feeds themselves on organic materials and dead fish that sink to the great depth [3].

Fig. 4.1 Roaring waves are charging toward the seashore along with whistling wind, composing a symphony of the sea with imposing grandeur. April 2009, Weihai, Shandong Province, Eastern China

In total, the oceans cover 71 % of the surface of the earth. Estimates say that the oceans contain nearly 1350 million km^2 of water; a number that there is hardly anything we can compare or associate with. The average depth is more than 3.7 km. The salt content is approximately 3.5 %. The world's largest ocean, the Pacific Ocean, has an average depth of nearly 4.2 km.

The Ocean surface is affected by the attraction from the sun and the moon, resulting in a rising and sinking surface level. When the sun and the moon are in the same lane (moon in line with the sun either between the sun and the earth or behind the earth), we get the highest tides.

Water evaporates from all water sources, but since oceans cover such a large part of the earth surface, the oceans become the most important source for evaporation and the moisture in the atmosphere. The evaporation is higher near the equator and is smaller toward the poles. When warm air rises, it expands and gets colder, and the water vapor condensates and forms clouds.

The sea currents transport warm water from the warm zones to the cold of the globe. An American hydrographer in the middle of the 1800s described the Gulf Stream like a river that never fails, even in dry seasons. The origin of the Gulf Stream is in the Mexican Gulf, and the mouth or the delta of this great underwater river is in the Arctic Sea. The Ocean currents are like rivers but many times the size of the Amazon River, and with at least the same ability to transport water, temperature, or objects that might be found in the ocean river.

The oceans are in more than one respect the most important stabilizer we have for life on the earth. In evaluating climate changes due to human activity and CO_2 emissions, the ability of the oceans to store CO_2 is an important factor in the estimations. According to Tim Flannery [4], the oceans have absorbed 48 % of all the carbon that comes from human activity in years 1800 and 1994, while the land has contributed to a net increase of the carbon in the atmosphere. The ability of the oceans of the world to absorb carbon varies quite a bit. The North Atlantic Ocean has relatively the greatest ability to absorb carbon. This part of the oceans represents only 15 % of the ocean's surface but has absorbed nearly ¼ of the carbon that has been emitted since 1800. The Arctic Sea has been an even stronger absorbent and has taken up 20 % of the emission.

Fig. 4.2 Gentle waves can also be characterized like this in the Silver Beach of Rushan, Weihai, Shandong Province, Eastern China in June 2012

Fig. 4.3 Infinity comes to reality in the meeting with the large oceans. You might imagine the curvature of the earth, and it becomes understandable when a ship comes up or disappears from the horizon

Fig. 4.4 Sea has many faces and colors at different time with different mood and can result in different mood. The color and mood of Hawaii sea is always so attractive, romantic, and full of imagination. December, 2010

Flannery explains this with a comparison picture from cold mineral water that keeps the gas longer than in the heat. Cold water has a greater ability to store carbon dioxide.

Flannery claims in his book *The Weather Makers* that the western part of the tropical Pacific Ocean is the warmest of the oceans. Between 1945 and1955, the surface temperature was often below 19.2 °C, but after 1976 it has seldom been below 25 °C.

The large oceans have sizes nearly beyond imagination, and there is hardly any place that human beings might so easily feel loneliness as in the middle of the oceans. The exemption might possibly be in the middle of Times Square in New York, or in centre of Shanghai in rush hours, but that is still another feeing (Fig. 4.5).

Temperature wise the world oceans is mainly split up in three layers. The temperature in a surface layer of about a 100 m varies considerably; from near zero near the poles, to somewhere over 30° near equator. As we go deeper, and the light disappears, the temperature gradually decreases to the depth of approximately 1 km. On the deep oceans further down, the temperature is fairly stable and might go down to −0.5 °C because of the salt content. Most of this cold water is flowing from underwater currents from the Antarctic.

The sea bottom is hiding as large resources of minerals, oil, and gas as on land. Over the last hundred years new technology has been developed to utilize these resources in a rational way (Fig. 4.6).

The Atlantic Ocean is bending like a large S between Europe and America and has of the western world also been called "the dam" that parts "the old world" from "the new world." The first European that was able to cross over to the other side was the Icelander Leifur Eriksson early in the 1000s. Some Norwegian books call him Leiv and claim that he was a Norwegian Viking. On Iceland, however, people are independent, good mannered, very interested in history and proud, and they very distinctly deny that he was Norwegian. Probably, we might say that Leifur had Norwegian blood and other Norwegian habits. His father, Eirik Raude, was the one that in 982 discovered Greenland. He had to escape from Jæren in Rogaland in Norway, and later from Iceland, and his settlement in Greenland was not far from the American mainland either. Only 300 km across the sound

Atlantic Ocean

	AREA – km²	Deepest point - m
TOTALT	86 557 000	8 605
Artic Sea	9 485 000	5 450
The Caribbean	2 512 000	7 680
Mediterranean Ocean	2 510 000	5 121
Mexican Gulf	1 544 000	3 504
Hudson Bay	1 233 000	259
Norwegian Sea	575 000	661
Black Sea	508 000	2 245
Baltic Sea	382 000	460

Indian Ocean

	AREA– km²	Deepest point - m
TOTAL	73 427 000	7 125
Bengal Bay	2 172 000	4 500
Red Sea	453 000	3 040
Arabian (Persian) Gulf	238 000	73

Pacific Ocean

	AREAL– km²	Deepest point- m
TOTAL	166 241 000	10 920
South China Sea	2 172 000	5 514
Bering Sea	2 261 000	4 150
Okhotsk Sea	1 392 000	3 363
Japan Sea	1 013 000	3 743
East China Sea/ Yellow Sea	1 202 000	2717

Fig. 4.5 The world Oceans [5]

from where the Vikings had settled in southwest Greenland was the American continent. On Greenland, the Eirik Raude family and others kept a Nordic settlement for 500 years. The Viking settlement in America, which they called *Wineland*, was rather short in historic sense.

More permanent traffic and European plundering of the "new" tempting and resource rich continent took place after the Italian, Portuguese, Spanish (some even claim that he too was of Norwegian origin), Christopher Columbus, looking for a new route to Japan and other Asian countries, came ashore on islands in the Caribbean in 1492. As a consequence of Columbus' four trips across the Atlantic, we witnessed a strong Spanish colonialization of the American islands and mainland both in the north and the south. This was soon followed by French and English activity. According to Thor Heyerdahl, with reference to several other experts [6], Columbus had in his younger days most probably visited Iceland, or Ultima Thule, as they claim that the island was called at that time. It is also

Fig. 4.6 "Artist view", Arco Barge. The barge, or rather, the LNG gas terminal, was designed and built in concrete in a dock in Tacoma, Washington, USA in 1973. For the construction 25 000 tons of concrete was used. Length 140.5, 41.5 m, depth 17.2 m. After birthing, the terminal was towed 16 000 km over the Pacific Ocean to the Java Sea in Indonesia in 100 days. The terminal was utilized as storing- and loading-terminal until 2006, when the field was empty

reasonable to believe that he knew the stories about the Viking settlements in Greenland, and about the large country Wineland further west. Before Christopher Columbus contacted King Ferdinand and Queen Isabella of Spain, and in 1492 got support for an expedition across the Atlantic Sea, he claimed to have been preparing plans for the expedition for 14 years, or from 1478, the year he came back from Iceland. Before Columbus got King Ferdinand's blessing, he had been working on the king for 6 years, and it was not a small payment that he was supposed to receive; 2 million *meraverdi*, status as a Spanish Nobleman, title as Head Admiral over the world oceans, and Vice King and Governor for life over land and island that he might discover. Assurance that the honorable titles that he would get should also be inherited by his oldest son Diego, and the next first born son in generations was added to the list. In addition, he got three fully supplied ships with 120 men [6]. Christopher Columbus did not die as a poor man. But at the time after his first voyage across the Atlantic, the situation became turbulent, and on his last voyage home, he was chained as a prisoner. Many of the promises from before the expedition were not fulfilled (Fig. 4.7).

Fig. 4.7 The statue of Leifur Eiricsson in Reikjavik, Iceland—a gift from USA at the 1000 years anniversary for the discovery of America

Thor Heyerdahl tried with considerable luck and with his Ra I and Ra II—voyages in 1969 and 1970 to show how the Atlantic Ocean might have been crossed long before the Europeans (Leifur and Christopher) concerning and before the start of our written history.

The Atlantic Ocean including the secondary oceans covers 29 % of the total ocean surface, or 22 % of the total surface of earth, the name originates from the old Greek historian Herodot from about 450 BC.

The S-shaped ocean is divided by an S-shaped mountain ridge from Iceland in the north to approximately 58° south.

The "Blue Ribbon" of the Atlantic Ocean was an important media case in its time, but has got little attention for the past half hundred years or so. The origin is claimed to come from *Gordon Bleu*. This has nothing to do with cheese or anything else from the culinary menu, but was an award in a knight order. The fight about the record for passenger ships to sail fastest across the Atlantic Ocean is old, but the 4-ft high "blue ribbon" statuette was not founded by the British member of Parliament Harold K. Hales until 1935. In total, it is recorded 35 ships that have had the *Blue Ribbon*—25 British, 5 German, 3 from USA, and one each from Italy and France.

Today, the connection across "the dam" is much faster in many respects. The start for laying the first telecommunication cable across the Atlantic Ocean was on August 6, 1857. The first cable broke after the first 4 days of laying work. They tried again in June of the following year. The cable broke again but, on August 5, 1858, the first cable line was completed. 11 days later, the first message was sent across, but the cable died after 3 weeks in use. A new and more stable cable was placed in 1866 and, in 1915, the first voice message crossed the Atlantic Ocean. The fiber optic cables of today, however, give signals across the ocean 4000 times faster than with the first cables [7].

The Indian Ocean is the smallest of the three giant oceans and covers about 20 % of the surface of the earth. The ocean is dominated in the north by India, and it stretches down to 20° south, when it is renamed to the Arctic Ocean. The island group Indonesia and the island continent Australia are borders in the east together with the Pacific Ocean, and the 20° east is the border to the Atlantic Ocean in southwest. Many "smaller" islands are found in the circumference of the ocean, but also the fourth largest island in the world, Madagascar, is found in the Indian Ocean. Bordering to the ocean in the north we find the probably oldest civilization on Earth, Mesopotamia. So it is reasonable to believe that this ocean was the first that human beings were sailing on in oceangoing vessels. The reason for Thor Heyerdahl's Tigris' voyage in 20 weeks in 1977–78 was to "prove" how primitive straw ships could travel over these ocean stretches.

Iraq's Museum has a ceramic boat model that shows that the art of sailing goes back at least to 4000 BC. [8]. We also know about ocean transport already from the first Egyptian Dynasty about 3 000 years BC. The Norwegian weekend newspaper, A-Magasinet, with reference to the scientific American journal PNAS reports in January 2013 that gene researchers have found that 10 % of the gens from the Australian aborigines comes from Indian. They claim the genetic entrance happened 4 200 years ago. Therefore, it seems like the Indian Ocean has been conquered quite a long time ago [9].

The Indian Ocean is very rich in resources too, and 40 % of the world offshore oil production comes from this Ocean.

Vasco da Gama was the first European that sailed over this Ocean when he from 1497 to 1499 sailed from Lisbon in Portugal to India and back. Vasco, however, was not the first European that sailed into the ocean. Already in 1488, King John II had sent Captain Bartolommeo Dias on a scouting trip along the African coast, and he managed to round Cape the Good Hope and recorded that Africa continued further northeastward.

Vasco da Gama's voyage was a very determined one. The Portuguese Kings had for years been irritated over the fact that the trade with the Middle East, India, and other countries in Asia was monopolized by rulers in the Mediterranean, the republic of Venice and Egyptian harbors. With the Portuguese strength at sea, to find a sea route to India would mean a lot of increased income and profit for the royal family. Vasco da Gama was appointed leader of an expedition and he on July 8, 1497 left Lisbon with a fleet of 4 ships and 170 men on board. Of these, 2 ships and 55 men returned, something that indicates some of the hardship they had to face on the voyage. After a very hard voyages via Tenerife on the Canary Island and the Cape Verde Islands, and a stop on the west African coast, and 3 months trip out into the South Atlantic Ocean, they rounded Cape Good Hope in December. They sailed up along the East Africa coast, and after several stops they headed out into the ocean at Malindi, north of Mombasa in what today is Kenya in February 1498. On May 20, the same year they landed in India near the city that is Calicut or Kozhikode in Kerala in South India today. Vasco and his ships left Calicut on August 20, and started on the voyage home. They rounded the southern tip of Africa on March 20, the next year, and the first ship, *Berrio*, was back in Lisbon on July 10, 1499. Vasco da Gama himself did not arrive because of different side trips until August 29, but was celebrated as a hero.

Vasco da Gama was also the leader of a fourth armada to India in 1503, and he was in 1524 appointed by the king of Portugal to Governor of the Indian state that Portugal established in 1519. The history about the great explorer Vasco da Gama is not only full of impressive knowledge and hard work, but also full of anecdotes about cruelty from the Europeans side, which might have exceeded what they experienced from the Muslim and Hinduistic local rulers in protecting themselves from the intruders. The discovery of the sea route was the start of a long history for Portugal as a colonial power, not only in Africa but also in Asia [10].

The Pacific Ocean is by far the largest of the world oceans, and covers nearly half (46 %) of the earth surface. It got its name when Ferdinand Magellan from 1519 to 1522 made the first voyage completely around the world. In Portuguese, the ocean was called *Mar Pacifico*, or the quiet ocean. Magellan, however, never experienced the complete around the world voyage, but died in the Philippines in 1521, while Captain Juan Sebastian Elcano led the expedition across the Indian Ocean and back to Spain.

The Pacific Ocean had been discovered by European some years before Magellan sailed across it, when the Spanish explorer Vasco Nunez de Balboa crossed Middle America in Panama in 1513 and found an ocean that he called *The Southern Ocean (Mar del Sur)*. The vision about *The Southern Ocean* has stayed well, and even today, there are many that look on the life in the Pacific as the life in *The Southern Ocean*.

After Magellan, Mendana sailed in several expeditions out into the Pacific Ocean Peru. He found the Salomon Islands in Melanesia in 1567 and the Polynesian Island in 1595 [6].

Probably, the greatest of all the Pacific Ocean explorers was Captain James Cook. Cook was a British navigator, topographer, and Captain, and had made a detailed map of Newfoundland before has started working in the Pacific Ocean. On his three Pacific Ocean expeditions, 1768–1771, 1772–1775, and 1776–1779, he contributed both to discover new islands and a detailed mapping of the Pacific Ocean area, from being the first to sail around New Zealand in the Southwest to the North American coast in Northeast and Antarctic in the south [11].

When Thor Heyerdahl, with his crew on board *Kontiki* from April 28 to August 7, 1947, sailed or let them self float, first with the Humboldt current northward and then with the equatorial ocean current westward from Callao in Peru to Raroia on the Tuamoto Islands, this was a physical proof at a voyage across the sea from east to west in the Pacific ocean also long before any Europeans had even thought in modern times. His findings that supported an alternative of population of the Pacific island also from the east were many. The island population in Polynesia regarded the strong ocean current as a gigantic river and described east as upward and west as downward. The indices for Thor Heyerdahl's theories were strong and many, from the sweet potato and other plants to mythological leftovers. Tiki was the human good, or the sea god that came from the world of light, from the east.

The Pacific Ocean has many islands, island groups, and island nations. The largest island, and the world's second largest island New Guinea, is an in the island group of Melanesia in the western corner of the ocean. The islands in the Pacific Ocean are very different both in origin and in appearance. The islands of Hawaii, for example, have volcanic origin, partly with active volcanoes, and the islands have many mountains. Other islands might only be low coral islands.

The Volcanic ring in the Pacific Ocean has the world's largest collection of volcanoes, with several hundred active volcanoes.

While the Atlantic Ocean every year increases in size, the Pacific Ocean every year gets a little bit smaller due to the movement of the tectonic plates. The Atlantic Ocean increases in size with about 2–3 cm per year on three sides, something that makes out a bit more than half a square kilometer each year [12].

The Channels

There have been built a number of channels around the world to connect the various waterways. There is hardly any doubt, however, that the two most important to connect world oceans are the Suez Canal and the Panama Canal.

The Suez-Canal

The 168-km long Suez-canal stretches from the bottom of the Suez Bay in the Red Sea, to Port Said at the Mediterranean side in Egypt. The importance of the canal to shorten the sea voyage from Europe to Asia can hardly be underestimated even if the important has been reduced somewhat in modern times with more modern ships and technology making the voyage around Africa more acceptable.

The width of the canal varies from under 100 m to more than 160 m, and it has a depth of about 13 m. There are moorings on the side of the canal for every 125 m for safety reasons. The Suez canal is a sea level canal, but there might be tidal water variations of 65 cm in the north and 1.9 m in the south. The canal has been built for one-way traffic, but there are all together 80.5 km with double lanes or passage width [13].

The sides of the canal vary from a slope of 4:1 in the north to 3:1 in the south, due to variation in the soil conditions.

The ideas to build the first canals were many, old, and very variable. The history of the canal stretches a long way back in time. According to the canal company, Egypt was the first country to build a canal to ease shipping traffic between continents, and the history goes 4000 years back in time.

The inscription on the grave of Weni the older, that lived in the sixth Dynasty in the old kingdom (2407–2260 BC.) tells about Egyptian channel building, and the reasons for the construction; waterway for warships, transport of stones for monuments [13].

The first canals were between the Red Sea and River Nile. Silting up of the water ways, however, led to stops of the canal traffic, and new constructions in several periods. New openings took place several times:

- 1874 BC. by Pharaoh Senausert III
- 1310 BC. by Sity I
- 610 BC. by Necho II
- 510 BC. by the Persian King Darius I
- 285 BC. by Ptolemy II
- Year 117 by the Roman Emperor Trajan
- Year 646 by Emir El-Moemeneen

During the Napoleons campaign to Egypt in 1799 came the first ideas to build a modern canal. An important reason was to create a considerable trading problem for the British. Napoleon, however, was told that the Red Sea was 10 m higher than the Mediterranean Ocean, something that led to anxiousness that the Red Sea would be emptied into the Mediterranean.

Then in 1833, a group of French intellectuals arrived in Cairo to investigate the canal building possibilities. The Egyptian ruler, however, was not interested.

Real progress came first in 1854, when French Ferdinand Marie de Lesseps managed to get Egyptian authorities interested. In 1858 was then the first canal company established. The company was established in Egypt with predominantly Egyptian and French shareholders. The British government bought the Egyptian ownership in 1875.

After a number of challenges, the modern canal opened on the November 17, 1869, after a construction period of ten and a half years. It is claimed that Egyptian farmers were forced to draft to work at the canal bin a number of 20 000 every 10 months [13], and that the work was far from well paid.

The Suez Canal was nationalized by President Nasser in Egypt in 1956.

The Panama Canal

The 81.6 km long canal crosses over the narrow Panama between South and North America, and connects the Atlantic Ocean with the Pacific Ocean. The canal has a width of slightly over 90 m and a depth of 12.5 m. Six double locks take the boats up from sea level to the 427 km² large Gatun Lake on 25 above sea level.

The construction of the Panama Canal was a 32 years project [14] that was completed in 1914 (due to the Second World War, the official opening was not until 1920), and that absorbed nearly 5 million m³ of concrete. The construction work also had considerable effect on the development in the concrete industry, as for example, the development of concrete trucks, standardized specifications etc.

The idea of shortening the route between the Pacific Ocean and the Atlantic Ocean and creating a waterway at the most narrow part between North and South America is old. Already in 1534, King Carlos I of Spain ordered to carry out investigations. With the development after the finding of gold in California and increased the American interest, in 1868 President Grant of USA also ordered to make investigations.

Real progress did not come until the Geographical Society in Paris discussed the case in 1876. On March 20, 1878, a French company made an agreement with Columbia (Panama was then a province in Columbia) about building a canal with fullback right to Columbia after 100 years. On May 15, 1879, the canal congress in Paris discussed the issue with 14 different channel construction alternatives. These were after the discussion reduced to two: Nicaragua and Panama.

The first spade stick was set in the ground on January 1, 1888. The original plan was to build a canal without locks, but after some time the proposal came to build ten locks. Gustav Eiffel (the designer of the Eiffel Tower in Paris) was appointed to design the locks. At the most, in 1894, there were 19 000 workers employed during the French construction period. It was difficult to find enough labor force, so many were hired from the West Indian island, and in particular from Jamaica.

In the beginning of 1889, the French company had no more money left, and on May 15, 1889, all work had to stop. The liquidation of the company took place in 1894 with a number of consequences. The initiative taker and the leader, Charles de Lesseps, became through several court cases sentenced to long prison punishments. Most of the sentences were made in hospital where he died 89 years old on December 7, 1894.

A new canal company was established on October 20, 1894. After a while, it came out that even this company had too limited finances. The work on the canal started up again on December 9, 1894 and, in 1897, the work force had grown to 4 000 men.

In 1898, more than half the capital of the company was spent, and contacts was taken with USA. On December 2, 1899, a meeting was arranged with the President of USA, William McKinley. However, it took another 5 years before an agreement could be reached.

It was President Roosevelt in USA that took action and put in military mussels in the work in Canal Zone. An agreement was made both with the French and with the freedom movement in Panama. On November 3, 1903, Panama declared independence from Columbia. The so-called Hay-Bunau-Varilla-treaty (named after the French and the American negotiators) secured USA sovereignty over 10 miles (16 km) wide belt in Panama, 5 miles on each side of the where the canal should be located was established. The agreement was ratified by Panama on December 2, 1903 and by USA on February 23, 1904.

Construction with USA leadership then started on May 4, 1904.

Chief Engineer John F. Stevens, later also called "the genius of the Panama Canal" convinced Roosevelt that six locks had to be built. He informed Philander Knox who spoke in the Senate about the case on June 19, 1905. The lock issue was approved by the Senate on June 27, with 36 against 31 votes. The two most important arguments from Stevens were

- To dig a channel without locks that was 15 ft. (45 m) wide at the narrowest would result in landslides.
- Without locks the channels would not be completed until 1924, but with locks it could be finished in January 1914.

Stevens resigned from the Chief Engineer job on April 1, 1907, and Roosevelt appointed a military leadership with important authority. Roosevelt was the first President of USA that traveled abroad when he visited the Panama Canal in November 1906.

The immigration of workers to the Panama Canal was formidable, and the challenge to get food and living quarters not less. At the most, on March 26, 1913 it was recorded 44 733 workers, not including the sick and the once not present. Most extreme was the immigration from Barbados in 1907 involving 19 900 men, or 30 % – 40 % of the grown up male inhabitants on the island community.

The work force numbers tell about 1000 persons in May 1904, rising to 3500 in November in the same year and 17000 in November the next year. For example in 1907, there were 5293 workers from Spain, 1032 from Italy, 1101 from Greece, 6510 from Barbados, 2039 from Guadeloupe, 2224 from Martinique, but only 13 from Panama. From USA, there normally came about 5000.

Sizeable (Water)—Rich Island Nations

We have many Islands or nations in the world on all continents and in all oceans. But probably more than any other place in the world, we have in the South Pacific island—nations with limited areas on land—but the state includes very large water areas. The island nation of Kiribati, for example, stretches from 168° east to 152° west with an area about the size of Australia, while the land area is less than, for example, Beijing.

In addition, many of us have a bit mystified imagination about these islands. If this is because the Pacific Ocean is so big and it is far from our own homes, many films, and books from Tor Heyerdahl and others, romantic movies like *Blue Lagoon* or other, or a combination will probably vary. Under any circumstances, these are countries or nations that might be associated even more with water than land and earth (Fig. 4.8).

Some data from some of them:

Fiji:

- 322 islands (+more than 500 smaller islands), whereas 106 have people.

Fig. 4.8 Island states in the southern Pacific Ocean

Small states South eastern Pacific	Land-Area Square kilometer	Population 1000
Fiji	18 330	839
Kiribati	717	88
Marshall-islands	181	53
Micronesia	701	109
Nauru	21	13
Palau	497	20
Salomon - islands	28 370	477
Samoa	2 831	178
Tuvalu	25	11
Tonga	748	104
Vanuatu	12	212

- Water area 194 000 km^2
- Highest mountain 1324 m above sea level (Fig. 4.9).

Salomon Islands:

- 990 islands
- Largest island: Guadalcanal—5336 km^2
- Highest mountain: Popomanaseu—2381 m above sea level

Tonga:

- 172 islands
- Land area about as smallest county in Norway, Vestfold.
- Highest mountain—1 030 m above sea level (Fig. 4.10).

The Reefs

The coral reefs have by some people been called the rain forest of the oceans. We find coral reefs in all the great oceans and in most of the subsidiary oceans. But reefs with the most variation in species are found where the temperature is highest and where there has been no ice age period. That is why we find five–six times as many types of corals in the South China Sea, in the Andaman Sea and the north side of Australia than in, for example, the Caribbean Ocean, and over ten times as many as in the North Atlantic.

Coral reefs are of a number of types and structures formed by lime deposits from living organisms. The

Fig. 4.9 Village dancing a late night with the food fire in the Yasawa Islands, Fiji

Fig. 4.10 Tongalesians entertain American church leaders by the ocean

inhabitants of the reefs make out 25 % of all marine species, including over 4000 types of fish, 700 types of corals and thousands of other species of plants and organisms.

The coral animals or polyps are found in many variations, from the fish eating to the ones that is fed by algae. When they die, they leave behind the calcium skeleton we normally regard as the corals. However, there are both soft and hard corals. A coral colony consists of hundreds of animals called polyps. Many of them multiply by division or lay eggs when fed by algae forms larvae. The larvae swim or float around and find a reasonable place to settle. Over years, decenniums and millenniums the skeletons built by death corals are building up on top of each other to form the coral reefs.

Most of the coral reefs we normally see in pictures are localized on relatively shallow waters under 60 m depth and in typical tropical waters. However, coral types have been recorded down to 300 m depth and in cold waters as far north as outside Alaska.

The Great Barrier Reef has been an icon, and this is probably the name and the reef area that most often is mentioned when people talks about coral reefs. The reef is located outside the northeast coast of Australia, and it has recorded at least 350 different types of corals. *Great Barrier Reef Marine Park*, instituted in 1975, includes an area of nearly 300 000 km^2. In 1981, the reef was declared a world heritage area. The reef, in fact, includes 3000 single reefs, where the whole reef area is 245 000 km^2.

An alarming article [15] claims that the coral system in the area covered by the reef has been reduced from 28 to 13.8 %, in the last 27 years. Storms account for 48 % of the destruction, coral eating starfish contributes to 42 %, and 10 % of the destruction is due to bleaching (whitening) because of higher temperature in the water. The researches who made the study propose immediate actions in removing starfish. They claim that the main reason for the increase in starfish is increase in the runoff of salts from agriculture in Australia. The cyclic increase in starfish comes much more often than before. If the starfish is taken away, the corals will grow with 0.89 % very year, and the ability to self-repair of the reefs will be reasonable. Tim Flannery [4], claims that much of the coral deaths on the Great Barrier Reef is due to *El Nino*. In 1997–1998, the rain forests in Indonesia had fires to a much larger extent than before, and for months clouds and precipitation were characterized by a thick red smoke cloud, in which contained a considerable iron. Late in 1997, a red colored ocean was observed outside the coast of Sumatra. The color was due to the increase in small organisms that were feeding on the iron from the smoke cloud. This led to reef destruction in a magnitude that will take tens of years for the nature to repair. During the El Nino in 2002, an even greater change in the ocean was experienced;the sea temperature increased considerably. The bleaching phenomena in 2002 collected a heating basin of ½ million square kilometers at Great Barrier Reef, which affected 60 % of the reefs, and in some reefs destroyed 90 % of the corals.

On the western side of the Australian continent, we find another great coral reef, possibly equally spectacular—the Ningaloo Reef. The reef, which is the greatest coral reef on the western side of any continent. It is located near the small town of Exmouth with 2500 inhabitants, about 1270 km north of Perth in the state of Western Australia. On the reef of 260 km long, it has been recorded not less than 250 different types of corals and 500 types of fish. In March to June each year large tribes of the biggest fish in the world, upto 18 m long whale shark come into Ningaloo Reef for mating play. Giant mantas and turtles can be found the year around and are also part of a possible fantastic diving experience, and not at least playful humpback whales that are playing on both sides of a boat on a diving trip, visual experiences that will be fixed on your brain forever. When the light is away from the sun you might see the whales playing down in the clear water, while in the light against the sun their playfully jumps up above the water become a fantastic visual sight (Fig. 4.11).

Unfortunately, we have seen a number of alarming reports about the destruction of coral reefs, both on the more exotic reefs and reefs in northerly waters in the Atlantic. Some claim that half the reefs in the world are threatened by human activity. Important destruction reasons are amongst others:

- Dynamite fishing
- Bottom trawling
- Temperature changes
- Earthquakes

Dynamite fishing: A number of places in the world there has been recorded a barbaric fishing method using dynamite. Dynamite is dropped down into the sea, and after the explosion, fish is damaged by the pressure and floats to the surface. The result, however, is that also the corals are destroyed. I experienced myself such damages on coral reefs in diving on the coast of Tanzania, a bit north of Dar es Salaam. The area with Stag horn,corals were decimated to the bottom with a few centimeters long coral pieces on the sandy sea bottom, which were rolling on the bottom in pace with the currents and the waves. The constant movement in the bottom structure made new creation of corals nearly impossible.

Luckily, some areas in the world have found remedies to repair such damages. For example, on some islands in the Maldivian, they have got all the different types of fish back by using special artificial reef components of concrete. The artificial reef components calm the movement on the bottom. In the best examples, they experienced all the different types of fish getting back after 250 days. However, it will take generations before the corals grow through the concrete components and repair the reef to its original function.

Fig. 4.11 There are lots of lobsters on Ningaloo Reef, and in a night dive, the light from the diving torch show a myriad of green lights in the coral wall like a night sky with stars where no constellations can be found

Bottom trawling: Many places in the world, even in North Atlantic waters, recordings are found of destructions of coral reefs from trawl fishing. In Spain and several other countries that have had some their coast line destructed in this way, they have even constructed artificial concrete reefs with a design that makes trawling very difficult.

Temperature changes: Several places in the China Sea, the Indian Ocean and around the coast of Australia, so-called Whitening of the coral reefs has been experienced. Changes in the ocean temperature have exceeded the tolerance level of the corals. Marine biologists from a number of countries have issued a warning about this probable negative effect of climate change. These warning are particularly serious because the changes take place in the part of the oceans that have the highest numbers of species of corals, fish, and other organisms. At the same time, these areas have been particularly rich on fish, possibly the most fish rich area in the world, and they are also birthing areas of many important species of fish. Serious researchers have also recorded that some species are increasing because of the warmer waters. There seems to be reasonable agreement about the fact that there will be changes, but it will take many years before we can know for sure what the consequence of the changes will be. It is also somewhat unclear if it is the temperature increases itself or the increase uptake in the oceans of CO_2, or a combination of this, that is causing the changes.

Oceanic uptake of anthropogenic CO_2 has been found to result in gradual acidification of the ocean, yet this is a very long time process even to cause a definite change of pH of the surface waters, let alone to say the change of whole ocean water through the uptake of anthropogenic CO_2 as well as natural physical and biological processes.

The variation in opinions and the conclusions in the reports might indicate that the situations might be different from one area to the other.

Earthquakes: Corals are brittle organisms and structures, and earthquakes on the sea bottom can make the coral stems break. By visual experience we have seen the ghostlike sight of the tilted white coral forest of 2–4 m coral trees on the sea bottom on the ocean floor outside Sharm el Sheik in Egypt. The white, dead corals had broken at the bottom of the stem and lay slightly at skew leaning at each other. However, this was the nature's own forces, so it is less to be concerned about than human made destructions, and with "natural" destruction it more reasonable to believe that the nature will heal the destruction itself, but it will take time, probably decenniums before the healing process is complete.

Other reasons for caring about the reef destructions are oil spillage, chemical and mineral and organic runoffs, human activity as shipping activity. Also bad behaviors from divers on souvenir hunt have been recorded as destructive, but the volume has mainly been limited.

The increase adding of nutritious salts has in several places led to increased starfish flowering; these feed on the corals and in other areas on the seaweed. As mentioned under the Great Barrier Reef passage, this might be an important reason for coral death in some places.

The Fiords

In some parts of the world and in Norway in particular, the fiords prolong the ocean by stretching the long tentacles of the oceans long, and sometimes very long into the land massive. The word fiord is also one of the few words from the Norwegian language that has been absorbed into world languages.

The fiord in a Norwegian context is special and somewhat different from the many: river outlets, deltas, sounds, bays, and gulfs that are found around the world.

We also find fiords somewhat similar to the Norwegian fiords in other countries: on the Swedish west coast, Limfjorden in Denmark, the Murmansk fiord in Russia, and the fiords on Iceland and on Greenland. Somewhat further from Norway, we find fiords on the west coast of Scotland, on the southwest coast of Ireland, on the west coast of North America—from Alaska down to British Columbia in Canada—on the south coast of Chiles, and on the southern island in New Zealand.

But, there are few fiords that are as spectacular as the Norwegian, with high mountains, dropping steeply into the ocean and with "nearly" comparable depths and lengths as the Scoresby Sound on Greenland with its 350 km, that is supposed to be the longest fiord in the world. The Norwegian Sogn Fiord with 204 km length is number two on the list.

The magazine *National Geographic Traveler* has appointed the Norwegian coast to one of the most attractive traveling goals in the world, and in the years 2004 and in 2009, the Norwegian fiords were awarded the number one position in the ranking (Fig. 4.12).

But, the Norwegian fiords are special also because there are so many of them. However, Norwegians have a tendency to call also small inlets and bays for fiords. Even some inland lakes are called fiords (Norwegian—fjord) like Tyrifjorden and Randsfjorden.

The typical fiord is really an underwater valley formed with a U-shape from melting of glaciers. This is also the reason why many fiords have relative shallow threshold in the outer parts of the fiord, while it might be many 100 m deep further into the fiord. The threshold contributes to the fact that the conditions inside the fiord are much calmer than

Fig. 4.12 There is not easy to find a more spectacular fiord than the Geiranger Fiord. Even in an overcast weather, the view from the "Eagle bend" against the waterfall "seven sisters" is an attraction that gets the tourist cameras to click like a small orchestra

in the big ocean just outside. Another phenomenon with the deep threshold fiords is there might be relative high salt content in the bottom of the fiords. The limited water circulation at the bottom water might also lead to formation of hydrogen sulphide and bad water at the bottom. On example of a small threshold fiord is the Drammen fiord with a threshold at Svelvik (Fig. 4.13).

The geographic conditions, combined with emission from the industry, runoff from agriculture and as recipient from sanitary installation can make some fiords very vulnerable for pollution. Some fiords have because of pollution got restrictions with respect to catching of shellfish and fish. Examples of this are the South fiord in Hardanger on the west coast of Norway and the fiords in Grenland in the eastern part of the country.

It is claimed that at least 24 Norwegian fiords have unreasonable pollution and have to be cleaned. The cost for this has been estimated to be between 10 and 50 billion kroner (RMB) [16]. The appropriate government agency has prepared a manual for the classification of the environment quality of the fiords [17].

From the Ringdal Fiord at the city of Halden that is border fiord between Sweden and Norway in the southeast, to the Varanger Fiord that is border fiord between Russia and Norway in the north, Norway has nearly 1200 named fiords. The significance is a fiord differs somewhat, and there are also somewhat different opinions on how to measure the length of a fiord (Figs. 4.14, 4.15 and 4.16).

Fig. 4.13 The Drammen fiord is narrow at the threshold. There are only a few 100 m across from Hurum on the east side to the beautiful small town of Svelvik on the east side

Fig. 4.14 The harbor of the city of Halden in Norway with the Ringdal Fiord that leads out to the Oslo Fiord, and then further to the Atlantic Ocean. The land in the background is Sweden

FIORD	COUNTY	LENGTH(km)
Sognefjorden	Sogn og Fjordane	294
Hardangerfjorden	Hordaland	183
Trondheimsfjorden	Sør – og Nord - Trøndelag	126
Porsangerfjorden	Finnmark	123
Lyngen	Troms	121
Oslofjorden	Oslo, Akershus, Buskerud, Vestfold, Østfold	118
Kvænangen	Troms	117
Ullsfjorden	Troms	110
Nordfjord	Sogn og Fjordane	106
Varangerfjorden	Finnmark	95
Romsdalsfjorden	Møre og Romsdal	94
Boknafjorden	Rogaland	94

Fig. 4.15 The longest fiords in Norway [18]

Fig. 4.16 Fish is of great importance to Norway, hence the rescue system behind the fleet. The rescue vessel "Christiania" was built in 1894, and is claimed to be the ship that has saved most lives. The ship has been in private ownership for many years now.

Ocean Interests

The dominance on the world oceans has through history been a military strategic challenge of considerable dimension. Also a small country like Norway yet with sizable sea interests has made its dispositions to protect the long coast and the impressive sea area outside the coast.

The Norwegian sea and ocean interests, both economically and defense wise, are stretching over ocean areas several times the size of mainland Norway, and the Norwegian naval history is as old as the history of the country.

To protect Norway from the sea, King Håkon, with the nickname "the good" (930–960) organized the so-called peoples fleet, also in Norwegian called "Leidangen." The citizens had a duty or obligation to help each other if they were attacked from abroad. This systems had been mentioned in several national laws earlier, but was finally institutionalized in the national law of 1274. The country was divided into so-called skipsreider that all had to secure one ship with a crew. The coastal population should build ships and a house for the ship, and take care of the maintenance. The ships should also be manned, and food should be provided. The ships were open long ships where the size was decided from the number of oars on each side, and this also gave the nominations of the ships. A 20-seated ship had a crew of 90–100 men. Some larger ships with 30 pair of oars were about 50 m long and 6–7 m wide. Such a ship might have had a crew of 260 men. In 1277, King Magnus Lagabøter mention 279 men, but earlier, there might have been a few more [19].

To mobilize this so-called leidang, they had instituted the hilltop fire system. This was a signal system with fires from well visual hilltops. The fires were ready prepared with spruce wood well in advance and with dry firing up wood ready. In addition to the hilltop fires a bow and arrow message system were added.

Norway has through history been attacked from the sea several times.

- The Jomsvikings, which had their headquarter in Jomsborg in the Baltic Sea, attacked Norway and concord Earl Haakon in the battle at Hjørungavåg on the west coast in 986.
- In September 999 or 1000, the Norwegian King Olav Trygvasson with the royal ship Ormen Lange and a Norwegian fleet was concord at the battle at Svolder. The attackers, and the winning party was the Danish King Svein Tjugeskjegg, King Olav Skøt of Svearike (former Sweden), the Jomsvikings and Earl Eirik Håkonsson from the Trondheim area in Norway. After the battle, Norway was divided between Earl Erik and the Danish King. So far the historians seem to agree. Where Svolder is, however, seems to be disputed. King Olav Trygvasson was on his way home from a raid to Pommern (Poland), so the theories saying that Svolder is place in the western Baltic Sea seems reasonable. Local historians, however, has the opinion that Svolder is in Svalerødkilen, a small fiord on the border between Norway and Sweden (Fig. 4.17).
- In the 1300s, Karels from Russia raided Norway from the north down to the Vesterålen Islands.
- Napoleon's continental blockade and the British blockade of Norwegian harbors crated hunger in Norway in 1812–1813.
- The last time that Norway was attacked from the sea was the German attack at the start of the Second World War in 1940 (Fig. 4.18).

But Norway has also used the sea as an attack road:

- In the period from the 700s and far into the 1100s, Norwegian Chieftains and Kings made a number of raids on a number of countries and areas, amongst others— Scotland, Shetland Island, Orkney Islands, Hebrides Island, Faeroe Islands, Island of Man, Ireland, Iceland, and areas in the Baltic Sea. Even so late as in 1263, King Håkon Håkonsson mobilized the coastal army in a raid toward west and Scotland. In the Norwegian history is also recorded the Norwegian discovery of Greenland and fights against "skrellinger" in Wineland (America). On Iceland, however, they will claim that they were the ones that discovered Greenland and America.
- In the 1400s Norwegian went on raids northwestward to the White Sea.
- A Danish/Norwegian fleet of seven ships under Captain, later Admiral, Peter Wessel Tordenskjold, attacked and concured a Swedish fleet of 13 ships in Dynekilen on July 8, only about 10 km south of Svinesund (today with a beautiful and famous border bridge between Norway and Sweden). The battle has been done immortal from a Norwegian "children song" about the Norwegian war hero.

Use of boats and ships for protection or for attack areas or enemies seems to have been started several places in the world, possibly independent of each other.

Fig. 4.17 Svalerødkilen a December day. A Svolder alternative

Fig. 4.18 "Sail ships have been very important for the Norwegian development"

- Battle with boats are known from the Qin Dynasty (221BC.) in China, but the first regular navy units in China is not known until the Southern Song Dynasty (1127–1279).
- The Chola Dynasty in South India (300 BC. to year 1279) must have had a powerful navy and they attacked and conquered Sri Lanka and several other areas in southeast Asia, amongst other what is Myanmar or Burma in the 900s.
- The Egyptians, Greeks, and Romans had naval ships often manned by rowers and with a tactic to hit enemy vessels through collision.
- In the 1500s and 1600s, Spain and Portugal built a considerably forceful navy that became a key tool for land conquering and collection of wealth from several countries and continents.
- After the attack of the Spanish armada on England in 1588, England after some time developed itself to be a great sea power and was for several hundred years the leading naval force in the world.
- From the beginning of the 1600s Netherland developed to be a major force at sea, with a very powerful navy that had battles both with Spain and England.
- The French navy had important battles with their neighbors from the 1600s and 1700s.
- In the 1800s important developments took place for the war ships, with for example: iron mantling of the hulls, transition to metal ships, development in the immovability of the canons, the size of the ships, and toward the end of century we got submarines, which completely changed the picture.
- In our time, USA has the most powerful marine fleet in the world; the US Navy has a tonnage comparable to the total of the next 17 navy's on the list together, or about 70 % of the total navy's in the world [20].

Use of boats and ships for promoting culture, trade, and friendship is also of great importance in history, which can be showcased by the two significant events on voyage to the ocean happened in Chinese history.

As the first one, Master Jian Zhen's voyage to Japan is a well-known event in Tang Dynasty (618–907). Jian Zhen (688–763) was said an expert on the Buddhist precepts and a good doctor of traditional Chinese medicine. In 742, Jian Zhen together with some of his disciples and some artisans started off for Japan by sea upon the invitation of Japanese monks studying Buddhism in China as well as the wish of the Japanese Government. They made six attempts to cross the sea during the following more than 10 years, yet all the five times voyages ended in failure due either to governmental interference or natural disasters. Finally in 753 their sixth time of attempt succeeded after 12 years' persistence and effort.

Jian Zhen stayed in Japan for 10 years in charge of preaching Buddhism and establishing Buddhist precepts as "Grand Master of Transmitting the Light" decreed by the Japanese emperor. He also introduced traditional Chinese medicine and Buddhist architecture to Japan. This event has been regarded as great historical significance in promoting the exchange and development of the Sino-Japanese culture and friendship between peoples as well as spreading the Buddhism more widely to Japan and East Asia.

Another significant voyage is Zheng He's expedition for seven times to what the Chinese called "the Western Ocean." This is a well-known history of China in Ming Dynasty (1368–1644) from the year 1405 through 1433. Upon the order of Emperor Cheng Zu of the Ming Dynasty, Zheng He (1371–1433) led a giant fleet to the Western Sea (today's Southeast Asia), carrying a total of over 27,000 people including soldiers and sailors and large quantity of goods. The largest vessels, 133-m-long "treasure ships," had up to nine masts and could carry a thousand people. The fleet during the seven times voyage in 28 years navigated the wide sea area from Ryukyu Islands, the Philippine Islands and Maluku Sea to the Mozambican Channel and the coastal areas of South Africa, and reached over 30 countries of Southeast Asia, east Africa and Arabia. They developed mutual trade, exchanged culture and technologies, communicated traffic on the sea, and promoted social and economic development in such countries and areas, which initiated a feat in the history of navigation and regarded as an unprecedented great historical period in Chinese history of trade and cultural exchanges. Unfortunately most of the documents were damaged by the fires of the Forbidden City in Beijing in Ming Dynasty.

Gavin Menzies, a British author and retired submarine lieutenant commander, according to Wikipedia [21] has written controversial books published in 2002 to promote the claims that the Chinese sailed to America before Columbus, though they have been categorized by historians as pseudohistory. In his book 1421, *The Year China Discovered the World*, he asserts that the fleets of Chinese Admiral Zheng He visited the Americas 87 years earlier than European explorer Christopher Columbus in 1492, and that the same fleet circumnavigated the globe a century before the expedition of Ferdinand Magellan.

Menzies later in 2005 [22] admitted in light of the latest evidence that Zheng He was not the first to sail to America. "One of the mistakes I made in my book was to say that Zheng He did everything. He had a legacy. Most of the world had already been mapped by Kublai Khan's fleet," he said. The copies of Kublai Khan's maps was found at the U.S. Library of Congress clearly showing North America. Menzies said he believes the maps, which are currently being carbon-dated, are from the late thirteenth century.

"The new evidence is likely to generate as much controversy as the book," as commented by Tan Ta Sen, president of the International Zheng He Society, and he believes that new evidence is "opening doors" but needs to be further substantiated.

In China, eleventh of July is Maritime Day devoted to the memory of Zheng He's first voyage.

References

1. Internet 18.09.12: Marina Trench, http://en.wikipedia.org/wiki/Marina_Trench.
2. Verdens Atlas (World Atlas), N.W.Damm & Søn AS, Oslo, Norway, 2002.
3. Terje Avner: *Yrende liv på verdens største havdyp (Life on the deepest ocean depths in the world)*, Aftenposten, Oslo, Norway, 21.03.13.
4. Flannery Tim: *Værmakerne (The Weather Makers)*, H. Aschehoug & Co, Oslo,Norway, 2006.
5. Atlas of the World: The Times, London, England, 2004.
6. Heyerdahl Thor: *Skjebnemøte vest for havet (Destiny meeting west of the ocean)*. Gyldendal Norsk Forlag AS, Oslo, Norway, 1992.
7. Riseng Per Magnus: *Hva er tykt som en hageslange, tusenvis av kilometer langt og fullt av Facebookoppdateringer?* A – Magasinet, Oslo, Norway, 25. januar 2013.
8. Ralling Christopher og Heyerdal Thor: *Thor Heyerdal – eventyret og livsverker, (. The adventure and his work of life)* Gyldendal Norsk Forlag, Oslo, Norway, 1989.
9. A-magasinet: *Indere seilte til Australia (Indians sailed to Australia)*, Aftenposten, Oslo,Norway, 25. januar 2013.
10. Internet27.01.13:*Vascoda Gama*, http://en.wikipedia.org/wiki/Vasco_da_Gama.
11. Internet 30.01.13: *James Cook*, http://en.wikipedia.org/wiki/James_Cook.
12. Internet 21.01.13: *Pacific Ocean*, http://en.Wikipedia.org/wiki/Pacific_Ocean.
13. Internet 02.10.12: Suez Canal Authorithy: http://www.suezcanal.gov.eg/sc.aspx?show.
14. Internett 21.11.09: www.pancanal.com/eng/history.html.
15. Eilperin Juliet: *Australias koraller forsvinner (the corals in Australia disappears)*, Aftenposten, Oslo, Norway, 12.Oktober 2012 (fra Washington Post).
16. Kaarbø Agnar: *Giftrensing av norske fjorder koster 50 Mrd (the poison cleaning of the Norwegian fiords costs 50 Billion)*. Aftenposten, Oslo, Norway, 20.12.00.
17. SFT: *Veileder for klassifisering av miljøkvalitet i fjorder og kystfarvann (Manual for classification of the environmental quality in fiords and costal areas)*, SFT (The Government Pollution Agency), Oslo, Norway, February 2008.
18. Internet 11.03.13: *Liste over norske fjorder (list of Norwegian Fiords)*. http://no.wikipedia.org/wiki/liste_over_norske_fjorder.
19. Engdal Odd G.: *Norsk Marinehistorisk Atlas (Norwgian Marie history Atlas)*, Vigmostad & Bjørke AS, Bergen, Norway, 2006.
20. Internet 21.01.13: *Navy*, http://en.wikipedia.org/wiki/navy.
21. Internet March 2, 2014, http://en.wikipedia.org/wiki/Gavin_Menzies.
22. Sonia Kolesnikov-Jessop, Did Chinese beat out Columbus? China Daily: Saturday, June 25, 2005.

Rivers and Lakes, and a Bit More 5

Brooks and rivers are formed when the precipitation is greater than the evaporation, and when at the same time porous ground do not manage to absorb and store the water. In a number of countries there are great differences between the rainy season and the dry season, and we find rivers that dry out when the rain stops. In Arabian countries in the Middle East, a Wadi in the desert landscape is such a dry river. This can be a popular transport route when it is not a rainy season or really most of the year. In such places water is also often found in the form of oases or less permanent wells. Some such dry rivers might only experience water a few days a year or even not every year.

Depending on the water amount and the velocity of the river, its ability to bring with silt, sand, and stones will vary. In the same way, the erosion ability of the rivers has its variation.

The Roman engineer and architect Vitruvius (e.g., Marcus Vitruvius Pollio), was also a geographer, and had his opinions and theories about the rivers of the world. In his famous book *De Architectura,* from about 25 BC, he writes about the rivers of the world (Fig. 5.1).

Some claim that Vitruvius was a very conservative author, but it seems like he was an icon in his time. A number of Roman authors seem to stay away from items if it had been written by Vitruvius or to choose to cite him instead.

Dr. Morris Hickey Morgan has translated Vitruvius book; "Ten books on architecture," which was printed by Harvard University Press in 1914. In a foreword to the 1960—edition of the book [1]. Albert A. Howard says about Vituvius: "There have been some discussions about when Vitruvius lived, but most people think that he wrote the book toward the end of the period with Nero as Emperor." *(Nero Claudius Drusus Germaricus, born Lucius Domitius Ahenobarbas on 17th of December, 37, inherited the throne in year 50, but did not become Emperor until he was 17 years old, in 54. He took the title Imperator in 66. He died on 9th of June, 68 [2].)*

The researcher claims that Vitruvius in no way was a good author. He was obviously "too much engineer," has simple sentences and formalistic formulations that remind about some kind of contract language. At the end of each chapter, he summarizes what he thinks is most important. In his "ten Books" he gives advice to architects over a very wide field of topics, from general education to general town planning, about materials, and not at least all the building types and structures an architect should know about.

In a modern sense, Vitruvius is not "only" an architect, but also much more. When he in book VIII writes about water, he shows an enormous span in geography knowledge, and he forms theories that hardly would be a hundred percent accepted today. He claims about the winds that they originate and come from the cold side, north, that they are cold and dry. Southern winds are warm and wet. This has effects on the rivers, where the largest and most important come from the north. He exemplifies this with amongst others: Ganges and Indus in India that come from Caucasus. Eufrat and Tigris in Syria, Ontus in Asia, Dnepr, Bug and Don in Colchis, Rhone in Celtica, Timavo and Po in Italy, Tiber in Mauretania, and several others, that all runs from north to south. With respect to the Nile, the history is this: Dyris starts up in the Atlas Mountains and runs west to the Heptabolus Lake, where after it changes name, floats southwards and runs into a swamp area. This is the kingdom of Ethiopia, and it forms the rivers Astansoba, Astaboa, and several others, where after it runs through cataracts in the mountains, north between Elephantis and Syene, and the plains near Thebes to Egypt, where it is called the Nile. He leads his river proof of the Nile further by reference to the animal life. Vitruvius sayings are interesting as a way of

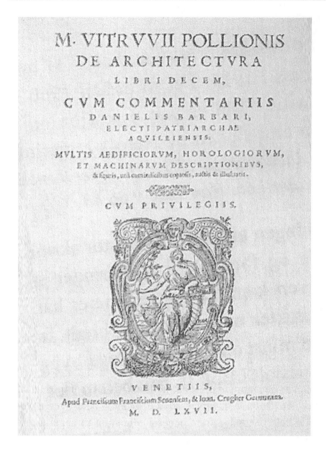

Fig. 5.1 The front page of Daniele Barbaro's famous edition of Vitruvius book from 1567

understanding the knowledge level people had 2000 years ago, even if modern climate specialists and geographers not always agree with all his writing.

All the great rivers have through years got their books, films, and television programs. For many of them, many books and traveling stories.

But, even if the great rivers and lakes are spectacular both in numbers and also to experience in reality, it is for many, possibly for most people that have not grown up near a big river or lake, the experience of the small brooks or rivers and the small ponds and lakes have made the strongest imprints in our mind. Thousands, not to say millions, of brooks and small lakes, might have given experiences and pleasure from a walking, bathing or fishing trip, or possibly from an overnight experience out in the nature. On the stomach down on the edge of a brook to stop the thirst or a small lake as a skating rink in winter, these might give just as strong feelings as the great river (Fig. 5.2).

The great rivers are many. My strongest mega-river experience was probably from a visit to Brazzaville some years ago. Brazzaville is the capital of the country which at that time was named Congo, previous French Congo. We were sitting at the edge of the swimming pool with a cooling drink in the hand on the overbooked hotel on a sunny Sunday. At the place we were sitting, we had the big, enormously big Congo River sailing lazily passed, which was less than a 100 m from us.

There was not much traffic on the river, but some primitive boats were sailing past, together with some branches of trees and some other wreckage wood from time to time. The hotel was full of refugees from the other side, mostly with Belgian background several generations ago. On the other side, we could barely see something we believed were some roofs of houses from Kinshasa, the capital in the big country, which at that time het Zaire, now the Republic of Congo, and earlier Belgian Congo.

Two capitals on each side of a big river, Africa's longest, 4700 km long, and also the world's deepest river. The river was nearly unbelievably wide. We cannot see clearly the houses on the other side, but from time to time we hear thunder-like noise from the other side. It is discussed if it is machine guns or bazookas. The time is so dramatic, and a civil war is going on in the country on the other side. Stories about escapes, suffering, rape, and massacres from the other side are told in the sun beside the peaceful and beautiful swimming pool. The impressive size of the Congo River does not only divide two countries and two capital, it divides between dramatic situations, cruelty and war, and a peaceful Sunday sun by the pool, an experience so contradictive and unbelievingly full that it is hard to imagine.

Another experience that we wish to share with you is the big contrast of the water quality observed when traveling the Yangtze River and Yellow River. The clear and friendly upper stream of Yellow River in Qinghai province was found to create a totally different image when the water travels downstream to the neighboring city—Lanzhou, capital city of Gansu province where the river containing more silts has become really yellow even with some offensive odor due to pollution along the river. Similarly the confluence of hazy Yangtze River and clear Jialing River also gives a remarkable contrast at Chaotianmen pier of Chongqing, a big city in southwest China. This reminds us of a Chinese idiom as 泾渭分明 [jīng wèi fēn míng] originated from ancient "Book of Songs," which actually is the description of confluence of the two rivers—Wei River and Jing River (branch rivers located near Xi'an, capital of Shaanxi province in the upper part of Yellow River) showing a remarkable difference in color due to the different content of silt. Later the idiom is often used as an analogy of absolutely different of character or attitude of people. These are right topics for children to understand the change of the environment and the awareness of protecting the environment.

Recordings of the length of the rivers might differ a bit between various statistics, as there might be variation in definitions, and the measuring might be somewhat difficult because

Fig. 5.2 River, brook or waterfall? Tioman Island, Malaysia. The island has been voted one of the world's most beautiful, and this brook or river is the place where the romantic "Bali Hai"—sequence was taken in the giant movie "South Pacific"

- It might be a matter of definition which side rivers and brooks that should be counted in.
- The origin of the rivers, in particular in tropical regions might vary from one season to the other and with precipitation from year to year.
- The mouth of the river, if there is any, might be a delta, where the transition to sea or lake might be difficult to define, and might change over time, as for example deposits from the river might make it longer over time.

In addition, lack of reliable maps makes the event difficult.

The length of the rivers in the world, therefore vary somewhat from one source to another.

Wikipedia has recorded 165 rivers with lengths above 1000 km. Of these, 17 are found in Europe, with the longest—Volga in sixteenth place and a length of 3645 km [3].

Many of the largest rivers in the world also have relations to more than one country—a source to discussions—both

River	Length (km)	Precipitation area (km²)	Medium water volume (m³/s)	Mouth	Continent
Nile-Kagera	6 650	3 349 000	5 100	Mediterranean	Africa
Amazonas – Ucayali-Apurimac	6 400	6 915 000	219 000	Atlantic	South America
Yangtze	6 300	1 800 000	31 900	East China Sea	Asia
Mississippi – Missouri-Jefferson	6 275	2 980 000	16 200	Mexican Gulf	North America
Jenisej-Angara-Selenga	5 539	2 580 000	19 600	Kara Sea	Asia
Huang He	5 464	745 000	2 110	Bohai Bai	Asia
Ob-Inysj	5 410	2 990 000	12 800	Ob Bay	Asia
Kongo-Iualaba-Lavua-Luapula-Chambeshi	4 700	3 680 000	41 800	Atlantic	Africa
Amur-Argun	4 444	1 855 000	11 400	Okhotsk Sea	Asia
Lena	4 400	2 490 000	17 100	Laptev Sea	Asia

Fig. 5.3 World's 10 longest rivers [3]

regarding origin and water rights (Figs. 5.3, 5.4, 5.5, 5.6, 5.7 and 5.8).

The world's by far the most volume rich river, the Amazonas River, has a precipitation area of 6 million square kilometer, that not only includes 60 % of the world's fifth largest country—Brazil, but also has influence from areas in French Guyana, Surinam, Guyana, Venezuela, Colombia, Ecuador, Peru and Bolivia. This area is bigger than Europe minus Russia.

Amazonas is by far, not the only river that has precipitation influence or runs through many countries. If we include the precipitation area and side rivers, the international complex gets even bigger. Nearly 50 countries on four continents are influenced by such "international precipitation areas or waterways," for example

- Danube runs through nine countries and its precipitation area comes from 15 countries: Albania, Bosnia and Herzegovina, Bulgaria, Italy, Croatia, Macedonia, Moldova, Montenegro, Poland, Romania, Serbia, Slovakia, Slovenia, Switzerland, Germany, Austria, and Ukraine.
- The Nile runs through nine African countries and its precipitation area in 11: Burundi, The Central African Republic, Egypt, Eritrea, Sudan, Tanzania, Kenya, Congo, Ethiopia, Rwanda, and Uganda.
- A number of countries are dependent on the water from Euphrates, Tigris and Jordan, amongst others: Egypt, Jordan, Israel, Lebanon, Turkey, Iran, Iraq, Syria, Kuwait and Saudi Arabia.
- La Plata has precipitation area from: Argentina, Bolivia, Brazil, Paraguay, and Uruguay.
  - Just to mention one example of important and large rivers, from different continents.

In addition, the "mega rivers"; Congo (13), Niger (11), The Rhine (9) and Zambezi (9), precipitation effects from 9 countries or more.

China is featured with a total length of 420 000 km long of various rivers and a total area of 56 500 km² of glaciers. Most of the large rivers have their source from the Qinghai–Tibet Plateau, and drop greatly from the source to the mouths. As a result, China is rich in water power resources, leading the world in hydropower potential, with reserves of 680 million kW. Asia's longest, and the world's third longest river, Yangtze, is probably the river in the world which has the most importance as a transport perspective,

Fig. 5.4 In the largest rivers we also find a number of rapids and waterfalls. One of the most famous is the Niagara fall on the border between Canada and USA with the world largest average annual flow of 2,400 m^3/s

Fig. 5.5 Sublime and misty Iguassu Falls with rainbow arch, shared by Argentina and Brazil, consists of about 275 falls along 2.7 km of the River, UNESCO World Heritage Sites. It is often compared with Southern Africa's Victoria Falls that separates Zambia and Zimbabwe

and together with the Yellow River, it is of uttermost importance for China, the world's most populated state, with rich history and culture and dynamic economic development (Fig. 5.9).

Yangtze, or Chang Jiang has its origin from glaciers on the Qinghai–Tibet plateau on 5000 m above sea level, and runs out into the East China Sea in a delta near Shanghai. Near Yibin in Sichuan Province, the river has come down to a level of 305 m above sea level, and from here and downwards, the river has a massive transport with river boats. Chongqing, where Yangtze meets the Jialing River we

are down to a level of 192 m above sea level. Chongqing is a city province with 30 million people, where about ¼ live in the city center. Chongqing is the economic center in southwest China with amongst others production of cars, motorcycles, chemicals, and electronics

On its way to the sea, Yangtze also passes Wuhan, one of the oldest cities in China and capital city of Hubei Province, with 10 million inhabitants and a history that can be dated back 3 500 years in time.

The Yangtze River drains 1/5 of the Chinese land area, and runs through 11 of the provinces of China: Qinghai,

Fig. 5.6 The Huangguoshu Waterfalls group comprising of 18 waterfalls with different features together with spectacular limestone karst landscape has been entitled in the world Guinness book of world records. It is located in 45 km southwest of Anshun, Guizhou province, southwest China. *Picture* Giant Huangguoshu Waterfall, Asia's largest waterfall. Water falling down 70 m off the cliff intensifies a loud thunder as thousands of drums which can be heard a few miles away. *Photo* Sui Hao

Fig. 5.7 But, small waterfalls might also be spectacular. This is the Seven Sisters in the Geiranger fiord on the west coast of Norway, running straight from the mountains and into the salt fiord

Tibet, Yunnan, Sichuan, Chongqing, Hubei, Hunan, Jiangxi, Anhui, Jiangsu, and Shanghai. Sixteen running-through cities include: Yibin, Luzhou, Chongquin, Wanzhou, Yichang, Jingzhou, Yueyang, Wuhan, Jiujiang, Anqing, Tongling, Wuhu, Nanjing, Zhenjiang, Nantong and Shanghai (Fig. 5.10).

The Yangtze River represents about half of the hydroelectric energy in China. We have in Chap. 7 told about the dam and power station at the Three Gorges, that is the largest hydropower dam in the world. In this chapter, we have told more about the river as a transport way.

The Yangtze River has a number of names and has changed over time, depending on the areas it runs through and the languages that are spoken in these areas. It is particularly below Yibin that the Yangtze name has been mostly used. The river is fed by a number of branch rivers in the

Fig. 5.8 The Amazonas river have a precipitation area of 6 million square kilometers

Fig. 5.9 The Yangtze River (left) and Yellow River (Right)

Fig. 5.10 River barge on its way down the Huangpu River, one of the delta rivers to Yangtze in Shanghai. The transport on Yangtze is annually more than 800 million tons

large area it runs through. After the river has run into the Hubei Province in a flatter landscape, the river also gets water from thousands of lakes. The largest of these lakes is Dongting lake, on the border between the Hunan and the Hubei provinces, and that receives water from most of the rivers in Hunan. In Wuhan, Yangtze gets its largest side river water volume, when it meets the Han River. Also the largest lake in China, Poyang, empties its water into Yangtze in the Jiangxi Province.

On its way to the sea, Yangtze gets water from more than 700 smaller rivers, where 10 of them has more water than 1000 m^3/s. Yangtze, every year leads more than 1000 billion m^3 of water to the sea. This is more than 17 times the water volume in China's second largest river, the Yellow River and more than 1/3 of all rivers in China [4]. It might be important to remember that more than 70 % of the water resources in China comes from the southwest of the country [5], where Yangtze and the Perl River are the most important recipients.

There have been built channels from Yangtze northwards, all the way to Beijing, and southwards to the Perl river delta.

Yangtze is not only an important resource and an essential part of the Chinese blood system, but it has also created challenges. In the last hundred years more than 300 000 people have lost their lives in floods in the river. In addition, it comes with the enormous material damages caused by the floods. Nevertheless, Yangtze has been in history consistently fond and proud of due to its spectacular scenery and rich historic culture. Here is an example;—a poem written by Li Bai in Tang Dynasty.

Leaving Baidi Town in Early Morning by Li Bai
In early morning I start when Baidi is amid rosy cloud,
Traveling a thousand li for Jiangling can return within a day;
As the riverbanks echo still with the monkey's cry aloud,
The swift boat has glided away myriad mountains.

Note

Baidi town—a scenery spot with a history of over 2000 years on the left bank of Yangtze River, which is located 8 km away to the east of Fengjie County and 451 km from Chongqing City, Southwest China
A thousand Li—here means a long trip; Li as a length unit in China is equal to 500 m in metric system

The Yellow River is the second longest river in China, but comes as number 3 after the Pearl River with respect to water volume. The river, originated from the Bayanhar Mountain Range in Qinghai Province and finally emptying into the Bohai Sea at Kenli of Shandong Province, is well over 5000 km long and runs through nine Chinese Provinces and Regions:—Qinghai, Sichuan, Gansu, Ningxia, Inner Mongolia, Shanxi, Shaanxi, Henan, and Shandong. The Yellow River in history is the birthplace of ancient Chinese culture and the cradle of Chinese Civilization.

The journey of the Yellow river toward the sea can be divided into three sections: the upper, the middle, and the lower section. The upper section goes from the start, passing through the Loess Plateau and to Inner Mongolia's Tuoketuo. The middle section goes further to the Henan Province and the lower section from there to the sea at Dongying, Shandong Province. In the middle section the river runs relatively steeply downwards in topography with rather lose soil. The soil is mixed into the river water and at the most more than 37 kg of sand is suspended into each m^3 of river water [4]. (Other sources mentions 34 and 36 kg [5]). As a comparison, the numbers for the lower part of the Colorado river in USA is 10 kg per m^3, and for the Nile 1 kg per m^3 [6]. This, the nearly frightening high amount of silt, makes the river very special and gives the reason for its name: The Yellow River.

When the river comes down to the plains in northern China, the velocity of the river is reduced, and its capability to keep the silt and sand fractions suspended is reduced. Consequently much of these materials are deposited on the river bottom. When the river reaches Kaifeng, 800 km from the sea, it contains up to 40 % of suspended materials. The river bottom might lift as much as 10–100 cm in a year, and the Yellow River's bottom is therefore lifted above the landscape around it. As the river bottom rises, the ditches around the river must be lifted and strengthened. In flood periods, the river has broken through the flood banks, something that obviously has led to many flood catastrophes. The history tells that over the last 2500 years, the river has broken through the river flood banks more than 1500 times, and has had 26 larger changes in its position [5], whereas six of them have been considerable. It is claimed that the river annually brings with it 1.6 billion tons of silt toward the sea and that 1.4 billion tons of the materials are deposited in the delta each year. These numbers are of such a magnitude that it is difficult to have a reasonable relationship to them. However, the annual amount is comparable to mountain of $1 \times 1 \times 1$ km (Fig. 5.11).

Floods are the nature catastrophes that annually lead to most deaths, and represents 40 % of all deaths in nature catastrophes. About 95 % of these deaths happen in developing countries.

In modern times, there are in particular three floods that are important for the Yellow River: the 1887—flood that took between 900 000 and 2 million human life, many from drowning and many others from hungers and epidemics. This flooding gave the river in principal the river location as it has today. The flood in 1931 took between 1 and 4 million human lives, and is the greatest natural disaster recorded in history in the world. During the war of Chinese against Japanese invasion, General Chiang Kai-Shek on 9th of June, 1938 tried in a desperate effort to stop the advancing Japanese army by making an artificial flood—breaking the flood banks of the Yellow River. Between 500 000 and 900 000 people died, but he did not manage to stop the advancing Japanese to conquer Wuhan, which at that time was the capital of China [8].

Fig. 5.11 The Yellow River—color print, where the artist have illustrated the trees on the head by signs that might be translated into; "*The Yellow River flows downwards as if it came from heaven*" and describes Li Bai's poem about the Tian Que-tower, in Chang'an, the old capital during the Tang Dynasty. Extract of this poem can be found in Chap. 2. The artist is Fu Jiayi (1944–2001). The artist has many pseudonyms [7]

Trying to put the flood disasters in the Chinese river in world perspective, we mention that Tim Flannery [9] claims that while 7 million people were annually hit by flood in the 1960s, and after the floods comes health problems. Calm water on flooded fields increases the amount of mosquitoes and malaria, yellow fever, etc.

As already mentioned, the Yellow River or Huang He, earlier called Hwang Ho, has many names. It is also called:—"mother of rivers," "China's mother" or "the cradle of Chinese civilization," and "China's pride," but also "The mud river," "China's grieve," and "The whip of the Han sons". Before the river enters the Chinese plains, it was in old Mongolian language called "Black river."

Along the Yellow River, it is not only featured with the breathtaking natural scenery of the river, but also characterized as a trip to explore the Chinese history and culture. Both beneficial and disastrous the Yellow River had been in the past, yet people in ancient China kept praising her, expressing their feelings, thinking and affections in virtue of her. The poem by Wang Zhihuan, Tang Dynasty is an example:

Beyond the Border
—Wang Zhihuan
Yellow River seems coming from white clouds afar,
The lonesome town is lost amid the steep mountains.
Why should the tunes of Tartar flute complain the willow,
The spring breeze never comes through the Jade Pass.

This poem pictures grand and vivid scenery: Overlooked the Yellow River to the up west, it seems to be flowing out of white clouds afar. In the steep mountains there stands a lonesome castle. Why should the flute tunes come along with nostalgic sentiments of willow (i.e., spring at hometown)? The spring breeze never comes through the Jade Pass, very close to the far west end of the Great Wall.

It has to be mentioned that more than one-third of the river system, based on a survey conducted by Yellow River Conservancy Commission in 2007, has water that is useless for aquaculture, industry, and even agriculture due to the fact that the river is classified in environmental class 5, the most polluted class. A considerable amount of the pollution comes from industrial activity and sewage discharge from the fast expanding cities [8].

Today the water in The Yellow River is getting better due to the considerable efforts that are taken to reduce the pollution, the water is strongly regulated through a number of dams and hydropower stations, whereas the seven largest produce 5600 MW of electricity (Fig. 5.12).

The Perl River, also named as Zhujiang, the third longest river in China and second largest by volume after the Yangtze, and the longest in south China, really consists of three rivers;—Xijang (west river), Dongjiang (east river), and Beijing (north river). Xijang is the longest of them and stretches out in a length of more than 2200 km. The river is located in a tropical zone with an annual precipitation of more than 1000 mm. This is the main reason for the large water volumes of the Pearl River (336 billion m^3) [5].

The Yarlungzangbo River, or Brahmaputra River in India, a transboundary river and one of the major rivers of Asia, which flows from its headsprings in the Tibet Autonomous Region of China first to the east and then south to India and further to Bangladesh converging with river Ganges and into the Indian Ocean through the Bay of Bengal, and boasts the Yarlungzangbo Grand Canyon, the largest canyon in the world, 504.6 km long and 6,009 m deep. The river with 2057 km long and over 5400 m drop in China's territory, is an important source for irrigation and transportation (Fig. 5.13).

The Lijiang River, the most beautiful river with spectacular limestone karst landscape in Guilin, Guangxi Zhuang Autonomous Region, China though it cannot compete with many other rivers in length. It belongs to the Pearl River

Fig. 5.12 The super speed train between Beijing and Shanghai crosses the Yellow River at a speed of more than 300 km/h, about halfway, near Ji'nan, The capital of the Shandong Province

Fig. 5.13 Bird view of the magnificent Yarlung Zangbujiang River, December, 2006. *Photo* Limin Yang

Fig. 5.14 Beautiful Guilin Landscape, China, a real water mountain picture. One early morning in October, 2010, Lijiang River

system with a total length of 437 km. The most spectacular part of 83 km is specifically called Lijiang (Fig. 5.14).

Hardly any mega river in the western world can contribute more to strong associations than the **Mississippi River**. Paul Robson's deep melancholic voice in Jerome Kern's *Ol' Man River* from the play *Show Boat* was one of the first tunes I heard, because it came from my father's favorite gramophone record and was played quite often. The deep bass was like coming from underneath the floor planks, and when he sang *lift that bale*, it was not difficult to imagine the shining sweat black workers at the bank of the gigantic river. Mark Twain's *Huckleberry Finn* and *Tom Sawyer* were not only persons from one of the favorite books from my childhood, but we played out the stories of Mark Twain near smaller Norwegian rivers, without thinking so much on the original river, the Mississippi. As we grew up we learned a bit about the great river and *"miss is sippi"* was remembered to manage to write the name correct. The river does not only represent the core in USA's culture, but also it has relationship to as many as 31 American states plus 2 Canadian provinces. The strong American cultural anchorage to the river is strengthened by William Faulkner's novels from the delta down near New Orleans, and the small myriad of composer, songwriters and singers that in one way or the other have contributed to the embracing to the river. Jazz from New Orleans and blues and gospel from

Fig. 5.15 Mardi Grass in New Orleans

Memphis are trademarks that can compete with the biggest trademarks in the world.

Mississippi is in many ways the "mother river" for USA, as Volga is for Russia and the Yellow River is for China (Fig. 5.15).

The more than 4000 km long river has its origin at 2700 m height in Brower's Spring in Montana, but most people claim that Mississippi really starts at the Itasca lake at 450 m height north in Minnesota. What is regarded as "upper Mississippi" goes from here and down to where Mississippi meets the great river Missouri in St. Louis. On this stretch, the river has not less than 43 dams, whereas 14 is located above the city of Minneapolis. Already when the river runs through "the twin cities" Minneapolis and St. Paul in Minnesota, the river has come down to a height of 209 m above sea level. It is not always easy to remember numbers, but it might be easier to remember that Mississippi from Minneapolis to New Orleans is about the distance from Beijing to Hainan or the length of Norway.

The middle part of the river is reckoned to be from St. Louis to where Mississippi meets the Ohio river in Cairo, Ohio, a distance of 290 km, where the river falls 67 m. The lower part of the river goes from here to where the river flows into the Mississippi delta in the Mexican Gulf, 160 km downstream from New Orleans.

The importance of Mississippi for USA, can hardly be underestimated. The river has created some of the most fertile land area in the USA and the precipitation area for the river covers close to 40 % of the country's land area. Other sources [10], claim that the precipitation area is about 5 million square kilometer or more than half of the land area in USA. The water volume is about 17 000 m^3/s in an annual average. The river really only crosses through two American states: Minnesota in the beginning, and Louisiana in the end, but it is border river for eight states: Wisconsin, Illinois, Kentucky, Tennessee and Mississippi on the east side and Iowa, Missouri and Arkansas on the west side. Mississippi gets supply of water from approximately 100 000 small and large rivers, where the four largest are Missouri, Ohio, Arkansas, and Red river (Fig. 5.16).

Mississippi has, of course as most large rivers, had its flood catastrophes, traffic accidents, and environmental problems. The down-silting has also led to that the river has changed several times its way downwards to the sea. River course changes have even led to the result that smaller areas

Fig. 5.16 Mississippi

have been placed on the other side of the river than the rest of the state in border states, because the river state line has been kept in the middle of the old river course

In the winter of 2012–2013, however, the river is not hit by flood but lack of water problem due to drought. New York Times reports on Christmas Eve in 2012 that the big river traffic in December–January between St. Louis and Cairo in Illinois that amounts to 7 billion USD is in danger [11]. The river has so little water that clearance of 9 ft. depth under the river barges is starting to be critical. It is possible that the traffic on Mississippi has to stop for a while. The US Army Corps of Engineers is working on improving the conditions, and is expected to make needed improvements in a few months. Another source [12] reports that the low water has exposed a number of historic objects on the previous river bed. Several old sunken steam boats have been visual, and an old Indian map engraved on a stone has caused the interest of the researchers. The map is expected to be 1200 years old.

An anthropology professor claims that it was not unusual in the 1800s that 100 steam boats passed St. Louis in one day. An average living time for the boats was 5 years, and he raised the opinion that there might be 500–700 such boats on the river bottom of Mississippi. Some of the boats were large, for example, the giant steamer *Montana*, that was as long as a football field. It went to the bottom after colliding with a floating tree in 1884 [13] (Fig. 5.17).

Europe's most central and second longest river after Volga, **Danube,** is not missing its special features as the only European river winding its 2845 km long way through ten countries: Germany, Austria, Hungary, Slovakia, Croatia, Serbia, Romania, Bulgaria, Moldova, and Ukraine. By the way, it is only for half a kilometer that the river runs through Moldova. It is also with some lack of precision that the river flows through 10 countries, for in fact it is a border river for two of them. In the annual New Year Concert televised to half of the world from Vienna, we are always reminded about the river through the Strauss wals *An blauen schønen Donau*, but the river is not much blue. We also doubt very much that it was blue on Strauss time either because the brown–gray river had its mud and lime content at that time too.

The Main-Donau—channel makes it possible to bring barges with cargo from the Black Sea to the Atlantic Ocean directly. In the Scandinavian and in the German speaking countries, we call the river Donau, while in English it becomes Danube, in Hungary the name is Duna, Serbers, Croates, and Bulgarians call the river Dunav, and the name is Dunaj in Slovakia. It is claimed that the name originates from Danubis in the Latin language—the Roman river God. Probably the name is even older and means to run or flow. Experts claim that the name is in family with the names of the Russian rivers: Donetsk, Dnepr, and Don (Fig. 5.18).

Human activity along the Danube River is probably some of the oldest in Europe and even in the whole world. It is claimed that one of our forefathers *Homo palaeohungaricus* settled in the Danube valley some for 250 000 years ago. Trade along the river can be traced 8 000 years back in time.

In Lepenski Vir in Serbia (1004 km from the mouth of Danube) a settlement was found in 1965, which is one of the oldest in Europe. In connection with the damming up in the "Iron Gate," the ruins of the settlement had to be moved a few tens of meters up the rived hillside.

The settlement was found by Professor Dragoslav Srejovic from University of Belgrade, and is dated back to

Fig. 5.17 Replica of a typical Mississippi—river boat is an obvious attraction Disney World in Orlando, Florida

Fig. 5.18 The longest European river, Danube is winding its way through 10 European countries. It is also a popular tourist attraction.

about 5600 years BC. The excavation indicates a settlement from 7000 years BC, with its heights between 5300 and 4800 BC. The settlement consists of a main village and 10 satellite villages. The findings from the excavation include tools of stone and bone, ruins of houses and religious objects and sculptures. The hut floor in one of the huts is claimed to be Europe's first use of concrete. They believe that the settlers in Lepinski Vir must have been some of the first hunting cultures in Europe from the end of the last ice age. Originally the village was located on a small horseshoe-shaped plateau near the Danube River, pushed in between cliffs in the great river. After the damming up of the Iron Gate, the whole place was moved 29.7 m up to avoid flooding from the artificial river–lake that is there today.

The village is a small settlement, possibly only for 4–5 families or for less than hundred inhabitants. The most important food source was probably fish. There is a clear plan for the village, and the burial place was provided underneath the buildings. The exemption, probably of religious reasons was a burial place behind the fire place in the house. All the houses have a distinct geometry, based on exact equal-sided triangles. Every house has a fire place in the shape of an oblong rectangle along the length axis of the house (Fig. 5.19).

The history of the Danube River is both manyfold and exiting, and the river was, during the Roman empire 2000 years ago, a kind of border river between the world's Roman center and the "barbarian" tribes and nations in the north and east. Reminiscence of Roman fortresses and settlements are still found in large amounts along the river, and only on the small distance between Vienna and Budapest there shall have been living quarters for 20,000 Roman soldiers. It is claimed that the Romans were the reason for the flowering of several of the capitals along the river today; *Vindobona* became Vienna, *Aquincum* became Budapest, and *Singidunum* became Belgrade.

A journey along the Danube river is not only a travel trough European history, but also a journey through

Fig. 5.19 Along the side of the Danube river at Lepenski Vir, we find one of the oldest settlements in Europe. The settlement has been moved 29.7 m up from its original position, and built into a small museum

Fig. 5.20 A journey along Danube also tells about wars and strong oppositions between ethnic groups of people through thousand of years—and also in recent times. In the Croatian harbor city Vokovar there will still in many years be visual examples and among people about the war between Serbs and Croates after the termination of Yugoslavia. Someone has, however, got the idea to humanize a war-hurt building with beautiful flowers

European architectural history; in palaces, castles, fortifications, bridges, etc. A journey along the Danube also gives an adventurely medley in southern European culture in general—everything from music to sculptures and food traditions and wine are natural experience objects as the river winds through nearly half of Europe. Only think of Snitzels from Vienna, Goulash from Hungary, Turkish inspired kitchen from the lower part of the river as kebab and yogurt from Bulgaria, or Serbian Shaslick. If this is not enough, you might be tempted by Rumenian redbeef soup in the lover river part, or tempting special German noodles in the upper part of the river. A complete menu of different types of beer and wines from the Danube countries are spanning over so many varieties of tastes that there should be options for everyone (Figs. 5.20 and 5.21).

Where Danube is border river between Romania in the north and Serbia in the south, the river runs through the so-called *Iron Gate*. Most of the Danube River runs through a relatively flat landscape, but in this area the river cuts its way through the southern parts of the Karpate Mountains and the Balkan Mountains. Four narrow gorges and several rapids made it impossible in the past for bouts to transport the river. The idea for the regulation is old, and already in 1831 the Hungarian authorities made plans for a regulation of the river. However, it was not until 1964 that this Rumanian–Yugoslavian megaproject that tamed the river started up. Two dams with hydroelectric power stations were built, *The Iron Gate I*—finished in 1972, 943 km from the mouth of Danube, and *Iron Gate II*—finished in 1984, 868 km from the Danube mouth. The dams have large locks and large river lakes on the upside. The increase of the water level up to 35 m led to that 17 000–23 000 people had to leave their homes and move to new homes. The dams obviously led to effects on the plant, animal, fish, and birdlife along the river. However, both Serbia and Romania have established considerable national parks on both sides of the river later on (Fig. 5.22).

Danube has a number of side rivers where Tisza or Tisa is the biggest. Tisza starts far up in the Karpatian mountains in Ukraine before it runs into Romania, and later flows over the Hungarian plains. Tisza flows into Danube in Serbia, not far above the capital Belgrade. Tisza has through times been one of the most flood-hit rivers in Europe, and has also in later years got a great amount of negative media attention due to very bad chemical pollution in the lower parts of the river. Flood regulation of Tisza on the Hungarian plains started already in the middle of the 1800s, and the river's original length through Hungary was reduced from 1419 to 966 km, plus a number of side channels and embankments. Many also recognize the river as the main venue through the famous Hungarian wine area in Tokay.

When Danube empties itself into the Black Sea, it forms a delta of about 70 × 70 km. There are not many roads in the delta but instead a great number of lakes—more than 400, and there is no lack of water ways. The main river is slit into three main water ways, where the northern one makes up the border between Ukraine and Romania, and the middle one is the one most important for boat traffic. The 120 km long northern water way in the delta is at places more than

Fig. 5.21 Water and sanitary water worker—in bronze—you might meet in the capital of Slovakia—Bratislava, 1869 km from the outlet of Danube in the Black Sea

Fig. 5.22 In the dammed up area, just above the dam at IRON GATE I, the river passes the serious face of Decubalus, a Drusic prins nearly 2000 years ago

Fig. 5.23 There are many churches and monasteries located along the Volga River. Here at Uglich, the river makes a sharp 90° turn before it takes journey toward south

1000 m wide, and through its many bends and beside many and large islands about two-third of the water in Danube is led further to the Black Sea. The very special condition in the delta gives the life conditions for more than 300 types of birds.

Europe's longest river, **Volga** starts at 225 m above sea level at the Valdai Mountains 320 km northwest of Moscow, and then runs eastwards until it gets water from the Moscow River, and then runs northwards. At Uglich it runs into the Rubinski Reservoir, once the world's largest artificial lake before it continues its journey southeastwards. At Kazan it turns southwards and later toward southwest, and before the river at Volgograd, previously Stalingrad, it makes a sharp turn and runs toward southeast to the Caspian Sea (Fig. 5.23).

Volga is the Russian "Mother River," both culturally and in an economical sense, 11 of Russia's 20 largest cities are found along the river. Even if Volga freezes up in nearly all its length during the winter season, the river is a very important transport way. The channel systems with large locks are constructed both up the Moscow River to Moscow, northwest from Uglich to St. Petersburg and the Baltic Sea, and northwards to the White Sea and the Arctic Ocean, and southwards to the Caspian Sea, and from Volgograd via Don to The Black Sea.

As other large rivers, Volga has been an inspiration for artists of most categories. The barge puller songs have been performed of the most influential orchestras and singer both national and international (Fig. 5.24).

Fig. 5.24 The barge pullers on Volga, a painting from the 1870s

The lower part of Volga is claimed to be the cradle for the Indo-European culture.

Smaller rivers/Norwegian Rivers;

Measured in length there is quite a distance from the mega-rivers in the world down to the Norwegian rivers.

There are about 4 000 Waterways in Norway, and nearly 3 000 of them are longer than 10 km, or have lakes larger than 1 km^2 [14].

It is easy to be fascinated by the great rivers of the world, but we must not forget all the others. The small brooks and rivers are normally not written so much about. However, the importance of them either as drinking water sources or in a recreation or in an energy context cannot be underestimated. The little brook that might start high up in the mountains and that grows bigger and bigger toward the ocean, the little brook that might have its origin in a small lake or a moor in the lowland has been both for drinking enjoyment and passing irritation—an asset for hiking people. Both the small and the large rivers have through the history had a deciding importance to all the people along the rivers. Rivers and lakes have been giving drinking water, important food, and income through fishing. The rivers have been the transport way for timber and an energy source for saw mills (Figs. 5.25 and 5.26).

As a small example of the importance of the many small rivers and brooks in today's society, we mention that the Norwegian association for small power stations on their internet page [15], tells NVE (Norwegian Waterway and Electricity Directorate) in 2011 was handling 63 cases of mini power stations, and that the waiting list at the end of the year was 600 cases for small power stations. When we know that the cost range for the small power stations range from about 10 million to more than 150 million NOK/RMB, these projects will have a lot of importance to small local societies in addition to the income from the power production.

Fresh water fishing in rivers and lakes is considerable. Annually 3 000–4 000 tons of inland fish are caught each year in Norway. The importance of the inland fishing is not new. From the Lake Tesse in Lom in middle Norway, more than 1000 year old fishing equipment has been found. It is told that Kong Olav The Saint about thousand years ago gave a local farmer Lake Tesse, if he agreed to be christened and to built a church [4]. The thousands of small rivers also have an immense recreation power even if the industry possibilities and the food-oriented fishing have been modernized away (Figs. 5.27, 5.28, 5.29 and 5.30).

Fig. 5.25 A somewhat symbolic picture from the town of Hønefoss (foss = waterfall) (about 60 km northwest of Oslo, Norway). The town was built around the waterfall in the Begna river, just before Begna runs into the Rand river—because this was important for the sawmills based on the timber transported in the rivers. The sculptor Knut Steen's monument "Oppgangssaga" (the saw) is located in the middle of the dry waterfall. Today, the water is utilized for electricity production, and is only let out in the waterfall in the tourist season. Times are changing. The original bases for the town of Hønefoss are gone, but it has at least made a monument, still if it is dry most of the year

Fig. 5.26 Through Norway's 4 largest inland city—Kongsberg, runs the river Numedalslågen. The river and the waterfall that divides the town in two is a nice and beautiful part of the town environment, even if the river and the waterfall did not have the same historic significance for this "silver mine town" as in Hønefoss

Fig. 5.27 The Asker River—a small local river, a day in October. The bridge is for the recreation walking path at Vøien, Asker, Norway

A very important ingredient in the Norwegian river picture is the waterfalls. In Norway there are often a very short way from the mountains to the sea. That is why Norwegian waterfalls are special. According to the book *Norsk Naturarv (Norwegian nature heritage)* [17], the University in Oslo has estimated that Norway has 9 of the 20 highest waterfalls in the world. However, there are only two that are not related to hydroelectric power stations (foss is the Norwegian word for a waterfall); Vedalsfoss in the community of Eidsfjord (645 m) and Rjoandefoss in the Flåm Valley in the community of Aurland with 563 m height. The book also mentions Kjeldsfossen in Gudvangen of 840 m height. The water volume is, however, so small that some do not regard it as a real waterfall.

The longest river in Norway, Glomma, also has its waterfalls, and the electric power stations in the river accounts for 10 % of the national annual power production. Glomma's origin is a few small lakes in the county of Sør-Trøndelag, where after the river runs through the counties of Sør-Trøndelag, Hedmark, Akershus and Østfold—before it empties out into the sea and Oslo Fiord at the town of Fredrikstad. About 1 % of the precipitation area of Glomma, however, comes from Sweden.

There were floated timber in Glomma up to the year of 1985, and the timber floating was at its peak in 1952, when 14 million timber logs were recorded at one of the timber stations [18].

The Glomma River has 17 electric power stations (Fig. 5.31).

Lakes

The origin for or the bases for **lakes** might be different;

- Sinking in the crust of the earth
- Dams from volcanoes or lava streams and earthquakes
- Dams from moraines in previously ice covered area
- Artificial dams from construction of hydroelectric power stations
- Along the coast, currents through deposits of sand over time might form lakes.

Fig. 5.28 Norway's longest rivers [16]

River	Length (km)	Precipitation area (km^2)
Glomma	619	41 857
Deatnu (Tana)	361	16 350
Pasvik river	360	18 510
Gudbranddalslågen/Vorma	358	17 548
Numedalslågen	356	5 554
Drammensvassdraget	308	17 113
Skiensvassdraget	273	10 811
Hallingdalsvassdraget / Snarumselva	258	5 253
Otra	245	3 750
Alta	240	7 389

Fig. 5.29 Numedalslågen is Norway's 5th longest river, but it is not among the most water rich. In the summer 2007, the river had a major flooding, that created major damages in some areas. The water level increased in some areas with close to 5 m. The picture is from Brufoss just below the village of Hvittingsfoss, where the river is at its most narrow. From the sea and up to Hvittingfoss, the Numedalslågen river is found rich resources of trout and salmon. The picture is taken just before the spring flooding in the Easter of 2013

The statistics showing the sizes of the lakes might change over time due to:

- Evaporation from shallow lakes
- Sedimentation of materials transported to the lake, or organic materials produced in the lake

There are lakes in the world without outlets, and where the evaporation is greater than the precipitation, resulting in salt water in the lake. Part of the precipitation area for such lakes, might be other areas than the lake itself, otherwise it would dry out. While the Genesaret Sea and Jordan are fairly fresh, the Dead Sea has a salt content of about 30 %, which is ten times higher than regular salinity of ocean water. Volga and Ural have fresh water, while the outlet lacking Caspian Sea has a salt content of 4 %. Titicaca in South America is fresh, while the outlet river ends up in salty lakes [19].

In Norway, there are recorded about 450 000 lakes of various sizes, which cover about 1/20 of the country. Eighteen lakes are larger than 50 km^2. A landscape with such a great number of lakes, are only found where the land area once have been covered by glaciers. The book *Elver og vann (rivers and lakes)* [14] claim that where this also is the

Fig. 5.30 A typical river valley. The picture could have been taken many places in the world, and Norwegian friends have wondered which Norwegian valley this is. However, the picture is from Tibet, China, about a quarter of an hour flying time southeast of Lhasa, and the river bottom is well above 3000 m height above sea level

case is in Alaska, parts of Canada and in the southern parts of New Zealand (Fig. 5.32).

Lake Mjøsa, located 123 m above sea level, is at the deepest 450 m. This means that the bottom is well below sea level. About 60 000 people do daily get their drinking water from Lake Mjøsa. The lake is special in several ways. Amongst others it is the only one of the greater lakes in Norway that has "harbor towns." The area around the lake is attractive to people, and amongst others is manifested through popular songs about the lake, etc (Fig. 5.33).

Lake Mjøsa is rich on fish of many attractive types, but due to the runoff from the agriculture in the important agricultural areas around the lake, the fish quantity has been reduced. In the last 25 years, however, the water quality has been improved again (Fig. 5.34).

Many of the Norwegian lakes are relatively deep, and the four deepest are also the deepest in Europe [20].

The recordings of the largest and deepest lakes in the world vary somewhat from one source to the other (Fig. 5.35).

The recordings for the largest lakes or inland seas change somewhat over time. One of the most dramatic and alarming changes we find is in the Aral Lake. This lake was for a long time the fourth largest in the world with an area of 63 800 km^2 [19] (Fig. 5.36).

The lake is located on the border between Kazakhstan and Turkmenistan, east of the Caspian Sea, and has no outlet. The rivers running into the lake Amu-Darja and Syr-Darja are intensely used as sources for artificial irrigation, and the reduced water volumes to the lake combined with evaporation have resulted in a situation where the shallow Aral Lake sunk with 15 m, now this great lake has been reduced to three smaller lakes. Several commentators have described the situation as the largest environmental catastrophe ever, and there have been worries that the lake might dry up completely. Kazakhstan has instituted several important actions in their part of the lake, amongst others a dam project that was completed in 2005. After the completion of this project the water has risen with several meters again. Simultaneously the salt content is reduced and the fish situation has improved so much that controlled fishing again can take place.

Fig. 5.31 Glomma at the end of March—just above Kykkelsrud Power station in the county of Østfold. This part of the river has long settling traditions. Ruins from a 10,000 year old Stone Age settlement has been found nearby, and also ruins of an old village fortress from the Viking era has been found

Fig. 5.32 The largest lakes in Norway [20]

	km²
Mjøsa	369
Røssvatnet	219
Femunden	203
Randsfjorden	140
Tyrifjorden	137
Snåsavatnet	122

Fig. 5.33 Lake Mjøsa, a summer day, with low hills and attractive islands that in many ways might remind of a coastal area

A similar lake dramatically drying out and decreasing in size is found in Lake Chad in Africa. The original lake was border lake between Niger, Chad, Nigeria and Cameroon. What is left of the lake now only has borders to Chad and Cameroon. The lake is now only 1/10 of what it was 40 years ago. From 1966 to 1975 the size of the lake was reduced by one-third mainly because of draught, lack of rain, and water from rivers. The additional reduction is human made. Toward the end of the 1980s, the water consumption from the lake to irrigation purposes increased considerably without precautions for the water coming into the lake. In addition, there comes a dam project that reduces the water flow to the sea. Lack of holistic and sustainable planning has not only reduced the size of the lake, but also has made catastrophic consequences for fishing and profitable agricultures in the area as well.

The deepest lake in the world, Lake Baikal, is located in Siberia in Russia on the border to Mongolia. The 635 km long and at its widest 79 km wide lake contains about 20 % of the nonfrozen surface water reservoir in the world. The lake is of course on the UNESCO's world heritage list from 1996. This is not at least because of its more than 1700 different plants and animal species, whereas two-third is not found in any other places in the world [21]. "Non-frozen" is by the way an imprecise expression for the lake that is located 455 m above sea level, and is normally covered with ice from January until May. The lake is found in a dislocation zone with mountains on both sides of the long lake. Recordings show that the dislocation is extended by about 2 cm each year. Lake Baikal is fed by not less than 330 incoming rivers. Only one river leads water out of the lake, the Angara River (Fig. 5.37).

Fig. 5.34 Rivers and lakes are both transport ways and obstacles. Norway's largest lake, Mjøsa previously had a ferry connection from the west to the east side. In 1987, a bridge took over the traffic from the ferry. The bridge became the first long bridge in the world with so-called high strength concrete

Fig. 5.35 Norway's deepest lakes [20]

Lake	Deepest depth -meter
Hornindalsvatnet	514
Salsvatnet	482
Tinnsjø	460
Mjøsa	453
Fyresvatnet	377
Suldalsvatnet	376
Bandak	325
Lundevatn	314
Storsjøen	309
Totak	306
Tyrifjorden	288

In a visiting report given by Siv Sæveraas in the Norwegian newspaper Aftenposten in January 2013 [22], it is claimed that the lake has more to offer than deserted beaches, crystal clear water, and fascinating nature. Mind cleaning and shamanism are the main points from the author. She visited the largest and most beautiful island in Lake Baikal, the 71 km long and 21 km wide Olkon Island. She feels like a lonesome tourist among the 1500 inhabitants on the island. Even if the Russians are in majority of the population along the Lake Baikal, the area is also characterized by the Mongolian ethnic group, the Burjats. The Burjats worship shamanisms, lamaistic Buddhism, and spiritual religions. They are of the opinion that Lake Baikal offers holy water, and a washing of hands and feet will increase your life with 5 years, a dip with your head will give you 10 years extra, and dipping your whole body will increase your life with 25 years. These figures should be a real magnet for tourists! Maybe it will not be so lonesome for visitors in the future.

The Lake Tanganyika is also one of the largest in the world, and the second deepest in the world. The lake has borders against four countries: Tanzania, The Republic of Congo, Burundi, and Zambia. The lake empties further into the enormous Congo River. It has been recorded 250 types

Fig. 5.36 The world largest lakes [19]

Lake	Area km²	Deepest depth in meter
Caspian Sea	394 299	946
Lake Superior	82 414	406
Lake Victoria	69 485	82
Lake Huron	59 596	229
Lake Michigan	58 016	281
Aral Lake	33 600	68
Tanganyika	32 893	1 435
Great Bear Lake	31 080	82
Baikal	31 500	1 741
Malawi Lake / Nyasa	30 044	706

Fig. 5.37 End of May. From 11 000 m height we see the snow clad tops in the mountains south of the Baikal Lake

of fish in the lake. The lake is located just south of equator and with its fantastic depth of nearly 1½ km, the water is divided into layers, with the warmest water on top. Typically for such lakes, it is difficult for the oxygen in the upper layers to blend with the water further down, while the lowest layers are more nutritious. The recorded climatic change has increased the problem of blending between the water layers. The monsoon is no longer strong enough to mix the layers sufficiently. Flannery claims that plankton in the upper layers is reduced to one-third in comparison with 25 years ago [9]. Researchers warn that if this continues, we can have a breakdown of one of the most interesting and many-fold ecosystems in the world (Fig. 5.38).

Lake distribution in China is very uneven with a concentration mainly in the areas of Qing–Tibetan plateau and the plains along the middle–lower part of Yangtze River. Fresh water lakes covers an area of 360 000 km², accounting for 45 % of China's total lake area, among which Poyang lake located north of Jiangxi Province and south bank of Yangtze River is the largest lake, covering an area of 3583 km². Salt lakes in China take the main part of 55 % of the total lake area, Qinghai Lake among which is the largest covering an area of 4583 km² with its water surface 3196 m high above sea level. Besides these, there are 286 000 artificial lakes and reservoirs in China (Figs. 5.39, 5.40, 5.41 and 5.42).

Fig. 5.38 A high bridge is crossing the Duero River in Portugal between the cities of Porto and Gaia. The many rivers in the world are a true gift and a great challenge to bridge engineers. Many of the bridges that have been designed and constructed have got great attention because of their elegance and engineering wise adjustment to acceptable esthetic norms

Fig. 5.39 Qinghai Lake, China's largest salt lake, where the blue water and sky melt as one and form a beautiful contrast with the yellow cole flowers. August 2003. *Photo* He Sui

Fig. 5.40 Nam-Co Lake, meaning as Sky Lake, China's 2nd largest salt lake after Qinghai lake and also world's highest lake of 4718 m above the sea level. It is located in the middle of Tibet Autonomous Region of China. It is said Nam-Co is a world of blues that goes beyond a normal vocabulary. The glaciers far away mix together with white clouds. *Photo* Limin Yang

Fig. 5.41 Beautiful lake formed by the thawing of glaciers up the mountain behind mirrors the snow mountain in itself. Baishuihe or White Water River with background of Jade Dragon Snow Mountain, Lijiang, Yunnan Province, Southwest China, May 2005

Moors, Wet Areas, and Swamps

Wetlands, according to the Ramsar Convention on Wetlands, are areas of marsh, fen, peat land or water, whether natural or artificial, permanent, or temporary, with water that is static or flowing, fresh, brackish or salt, including areas of marine water the depth of which at low tide does not exceed 6 m. In general, wetlands are water bodies but also include land. Wetlands, as one of the three global ecosystems together with forests and oceans, exist in every country and in every climatic zone, from the polar regions to the tropics. They are so important and considered as the Kidney of the Earth, yet highly variable and sensitive, and have been integrated throughout the history of human survival and development. They can be vast storage of water and flood damper, important reserves for plant and animal species, and products and energy bases. However, the global wetlands cover only about 5.7 million square kilometer, equivalent to 6 % of the earth's land area.

Thirty percent of the fresh water on the earth is ground water. 11 % of the surface water is in swamps. Moors are really none of the two, neither ground water nor swamp, but at the same time a bit of both.

The moors act as swamps in the nature and are a very important regulator for the ground water. The wet areas are an extremely important resource for the many folds we have in our nature. Many of the species of animals, fish, and plants have all their life cycles in the wet areas, while other

Fig. 5.42 A close view of Baishuihe or White Water River creating a clear pond world comprising of water, sky and snow mountain, May 2005, Lijiang, Yunnan, China

types use the wet areas due to their richness of food substances in their search for food, or as resting places in their nomad voyages, or as their resting place before ocean voyages.

Permanent protection of wet areas is an important international environment action area. A wet area convention as **Convention on Wetlands of International Importance Especially as Waterfowl Habitat (i.e., the abovementioned Ramsar Convention on Wetlands)** was signed in February 1971 in Ramsar, Iran and enforced in December, 1975.

The Pantanal [23] is the world's largest tropical wetland area, and is located mostly within the Brazilian state of Mato Grosso do Sul, and extends into Mato Grosso and portions of Bolivia and Paraguay. It sprawls over an area estimated between 140,000 and 195,000 km^2. Various subregional ecosystems exist there, each with distinct hydrological, geological, and ecological characteristics.

Moors cover large areas, in fact 10 % of the Norwegian land area. However, about one-fourth of this has been ditched, so the real moor area in Norway covers a bit more than 20 000 km^2 [17]. The moor percentage differs quite a bit over the country. On the lowland in south Norway, the condition is most favorable, and we find the deepest moors. Here more than 7-m deep peat layers might be found. In previously small lakes that have been grown in, the moor might be even deeper.

Nordre Øyeren wet area not far and northeast of the Norwegian capital Oslo is one of the special wet area protected by the Norwegian government. This area got its protection status in 1995, and is part of the Glomma (Norway's longest river) waterway system. The area includes 63 km^2, whereas 55 km^2 are water, 20 km^2 are delta plains (including water) and 7 km^2 are island. T area has been recorded with 308 types of plants, 25 types of fish, and 260 types of birds [24, 25]. This wet area is the largest inland delta area in the Nordic countries in Europe [26] (Fig. 5.43).

Wetlands in China covers an area of about 113 000 km^2, ranking first in Asia. Sanjiang (Three Rivers) Plains and Qinghai–Tibetan Plateau among which are regions, where marsh wetlands are intensively distributed. There are 30 wetlands in China up to February 2005 which have been entitled in the list of important international wetlands. Examples include the Birds Island in Qinghai Lake, Xingkai Lake in Heilongjiang Province, far northeast province of China, etc. Unfortunately there has been a dramatic shrinkage of the wetlands due to unreasonable exploitation and even damage since the end of 1970s. What is good is that China joined the Wetlands Convention in 1992 and issued China Wetlands Protection Act Plan—after completing a 6-year national survey in 2000 and National Wetlands Protection Projects Programming in 2004. So far, 40 % of natural wetlands have been well preserved. Up to the year 2030, a target of properly protecting more than 90 % of natural wetlands resources will be completed (Fig. 5.44).

In the global context, wetlands, as one of the world's most important environmental assets and abundant resources of fish, fuel, and water are vulnerable to over-exploitation.

Fig. 5.43 The Nordre Øyern wet area is a popular hiking area, and on the entrance roads and paths, there is good information about what can be expected to be observed

Fig. 5.44 Red Intertidal Mudflat in October, located in Panjin, Liaoning Province, estuary of Liaohe river, north of Baohai Bay, China. It is one of the most beautiful wetlands in China and the world largest reed marshes covering over 20 km^2, where more than 236 species of birds dwell including very rare species of cranes. Suaeda salsa growing there creates a spectacular red beach every autumn in September to October. Thank to unique saline-alkali soil and distinct four seasons

The rate of loss and deterioration of wetlands is accelerating in all regions of the world. The pressure on wetlands is likely to intensify in the coming decades due to increased global demand for land and water, as well as climate change.

International Wetlands Standing Committee designated February 2 every year as World Wetlands Day.

References

1. Vitruvius: *The Ten books of Architecture*, translated by Morris Hickey Morgan, Dover publications, Inc., New York, 1960.
2. Internet 05.10.09: Nero, www.Wikipedia/.
3. Internet 16.09.12: http://no.wikipedia.org/wiki/Liste-over-verdens-lengste-elver.
4. Zheng Ping: *China's geography*, China Intercontinental Press, ISBN 7-5085-0914-5/K-751.
5. *China in Diagrams*, China Intercontinental Press, ISBN 7-5085-0842-4/K-726.
6. Internet 26.11.12: *YellowRiver*, Wikipedia, http://en.wikipedia.org/wiki/Yellow_River.
7. Tongbo Sui. Mail 06.12.12 with translation from a Yellow River- print.
8. Arneson Steinar: *Energi Lex 2001*, Villrose Norsk Forlag, Oslo, Norway, 2001.
9. Flannery Tim: *Værmakerne (The Weather Makers)*, H. Aschehoug & Co, Oslo, Norway, 2006.
10. Robinson Andrew: *Natursjokk (Nature Shock)*, Grøndal Dreyer Forlag, Oslo, Norway, 1995.
11. Schwartz John: *Cargo Continues Moving on Missisppi River, but Perhaps Not for Long*, New York Times, 2012.12.24.
12. Internet 04.01.13: *Mississippi River, other tributaries drained drought reveal sunken treasures.* http://blog.gulflife.com/mississippi-press-news/2012/12/mississippi_river_other-trib.
13. Internet 03.01.13: *Missisppi River*, http://en.Wikipedia.org/wiki/Mississippi_River.
14. Eie Jon Arne, Faugli Per Einar, Aabel Jens: *Elver og vann- Vern av norske vassdrag, (Rivers and lakes. Protection of Norwegian*

References

waterways) Grøndahl og Dreyers Forlag AS, Oslo, Norway, 1996.
15. Internet 13.10.12: Småkraftforeninga: *Konsesjonsbehandling, (Small powerstation association: Consessionhandlig)* http://kraftverk.net/visside.php?id=21.
16. Internet 16.09.12: Store Norske Leksikon*(Big Norwegian Encyclopedia):* http://snl.no/elv.
17. Hågvar Sigmund og Berntsen Bredo: *Norsk Naturarv (Norwegian Nature Heritage),* Andresen & Butenschøn AS, Oslo, Norway, 2001.
18. Internet 24.03.13: *Glomma,* http://no.wikipedia.org/wiki/glomma.
19. Internet 17.09.12: Large Lakes of the World, http://www.factmonster.com/ipka/A0001777.html.
20. Internet 17.09.12: Store Norske Leksikon: http://snl.no/innsj%C3%B8.
21. Internet 23.10.12: Wikipedia: *Lake Baikal,* http://en.wikipedia.org.wiki/lake_baikal.
22. Sæveraas Siv: *Du store Bajkal!(Large Lake Bajkal);* Aftenposten, Oslo, Norway, 5. January 2013.
23. Access on June 3, 2014. Pantanal: http://en.wikipedia.org/wiki/Pantanal.
24. Internet 30.03.13: *Nordre Øyern naturreservat, (Northern Øyern Nature area)* http://no.wikipedia.org/wiki/Nordre_%C3%98yern_naturreservat.
25. Kunnskapsforlaget/ Norges Naturvernforbund: *Natur- og miljøleksikon,* Hovedredaktør: Henning Even Larsen, Oslo, Norway, 1991.
26. Internet 30.03.13: *Miljøstatus i Oslo og Akerhus, (Environmental status Oslo and Akershus)* http://osloogakershus.miljostatus.no/msf_themepage.aspx?m=2483.

Transport Lanes

6

All the many intricate and sometimes peculiar water ways we have are essential to the blood system of the world's transport system, and they have been there much longer than our history we record. On the world oceans, we secure cargo and energy transport between the continents. Initiatives to control these possibilities have led to wars, explorations, and development, not to mention the cultural heritages buried in the songs, poems, and books about seamanship, hardship, and joys on the many oceans.

Rivers and lakes were probably the first maritime transport ways, and building of channels or canals in a large number of variable types to improve the systems has taken place on most continents. Canals as "city streets" still function reasonably well in a number of cities. Many of these cities have become tourist magnets possibly just because of their old and historic canals.

Most well known, at least in the western part of the world, is probably Venice in Italy, but rather similar canals as transport ways are important tourist attractions in a number of cities (Fig. 6.1).

Canala Grande, or **Grand Canal**, or "The Great Canal", is a canal name that not only exists in Venice, but has been in many places throughout the history. One of the oldest examples of "modern canals", is Grand Canal in Ireland. This canal was initiated already in 1754 and completed in 1804, and is the southern canal that connects the capital Dublin on the east side of Ireland via Tullamore with the Shannon River on the west side of the island. The canal has 43 locks and a length of 132 km. The idea for the canal came already in 1715, but it was not until 1757 that the Irish parliament approved the first financial support for the construction. In 1772, The Grand Canal Company was established to secure money for the difficult work that was left. The company even invited the famous British engineers John Smeaton (The designer of the Eddystone Lighthouse) and his assistant William Jessop to Ireland for 2 weeks so that they could give their advice. Even if the canal was not opened until 1804, it was part of the canal used for passenger traffic already in 1779. The channel is still in use, but the last commercial loading barge went through the channel in 1960 [1] (Figs. 6.2, 6.3, 6.4, 6.5 and 6.6).

The large rivers and the water systems between them are of uttermost importance in a number of countries and in the transport between countries. Yet sometimes rivers can also be the hindrance we have to get across using ferries and bridges in various types. Earliest suspension bridge as an example across the valley was said to be built in 519 in China's history, while similar bridge appeared a 1000 years later in the sixteenth century in the west. A modern example is the 11 bridges and 1 tunnel connecting the three parts of prosperous Wuhan city located on the both sides of Yangtze River in Hubei Province, China. There are moreover five bridges under construction and five under programming (Figs. 6.7, 6.8 and 6.9).

The rivers have also been an important transport way in Norway, and not at least until a few tens of years ago, the rivers were absolutely the most important transport possibility for timber to sawmill and the production of cellulose with the factories located downstream in the rivers (Figs. 6.10 and 6.11).

Another example of other canalization projects in Norway is the canalization of the Skien waterway that started in 1854.

We do not for sure know how old the tradition of floating of timber is as a transport possibility, but we know that registrations of timer floating in the Drammen waterway even existed before 1350. A law about timber floating in the rivers was instituted in 1854, and floating associations were started [32].

Timber floating on the river normally took place as loose single logs, but on the lakes, the logs were normally connected to floats. The floats were drawn by hand from land or handled by rowing boats.

On the rivers, most of the floating took place during the spring flood when there was a lot of water in the rivers. To

Fig. 6.1 Canal Grande, and typical tourist enjoyment in the gondola in Venice, Italy

Fig. 6.2 The canals in Nanjing in China have a colorful and romantic touch, and are a popular traveling goal. They are also attractive for a late evening meal. Nanjing is today a city with more than 5 million inhabitants, but already in the 1400s half a million people was living there, and the city was then the largest city in the world. Nanjing means the capital in the south and has been the capital of China in several dynasties. The first period was about 220–280 BC. The Qinhuai River and its canals were very important for the city

secure enough water floating dams were also built. In these dams, water was collected to be let out when the timber came down the river to the dam area.

The last timber floating in Norway is claimed to take place in the Telemark Canal in 2006, but the commercial floating in most rivers was terminated and transferred to trucks in the 1980s and 1990s.

The expression *Telemarks Canal* came into use toward the end of the 1900s. The first part of the canal was the Norsjø–Skien Canal that was built in the years 1854–1861. The background for the construction was to increase efficiency in the timber floating. When the single log was let into the river, then some logs might hang up and others might get lost as sinking timber. The timber might also be destroyed in the

Fig. 6.3 The canals in Amsterdam in Netherland is also a tourist magnet

Fig. 6.4 Gothenburg, Sweden has its popular canal boat "Paddan" as a sightseeing alternative

bunches that were made sometimes. With the canal and the locks at Skien and Løveid, the timber could be floated in floats and under more control on their way to the downstream sawmills at Ulefoss and in the town of Skien (Fig. 6.12).

Then in 1892, Bandak–Norsjø Canal was constructed. In European traveling handbooks, the canal has been called *the eighth wonder of the world*. In total in the Telemark Canal, there are eight lock sites with all together 18 lock chambers. The highest lift is at Vrangfoss, where five locks lift up a total of 23 m.

In August 2012, the Norwegian national television made a fantastic live broadcasted minute for minute trip through

Fig. 6.5 On and around the largest rivers in Southeast Asia, millions of people are living most of their lives on and from the rivers, on the side rivers and the canals in close connection to the main river. The city picture is swarming—from the Praya River, Bangkok, Thailand

Fig. 6.6 A tourist boat at the Moldau River (Vitaya) in Praha. The Kafka museum at the river bank behind the boat—and the Hradcamy castle at the hill in the background. Moldau runs into Elben 50 km further north. The 1091-km-long Elben is an important transport way in one of the most densely populated areas in Europe. Elben runs further toward northwest and passes on its way to the sea the cities of; Dresden, Magdeburg, and Hamburg, before it empties into the Arctic Ocean at the delta near Cuxhaven

the canal with two passenger boats. The passenger boats MS "Viktoria" and MS "Henrik Ibsen" are trafficking the canal. "Viktoria" has been in traffic since 1882 (Fig. 6.13).

But it is probably the coastal water ways and the ocean that in the past, today, and in the future that are the most important transport ways on water for:

Person Transport, Ferry Transport, and Cargo Transport.

Sea transport is a very large and many folds area in the society, so we are only able to give a limited glimpse into this part of our society.

In the national transport plan the Norwegian Coastal and Fishery Department amongst other say [2]:

Inland ships account for about 45 percent of the total transport, and sea transport is completely dominant when it comes to international transport. Still more cargoes have to go into ship transport if we shall reach the goals with increased mobility, increased safety and less emission to the environment.

Prognoses forward to 2040 show increased growth both in freight on ships and on trucks, but the road transport will increase more than sea transport.

The statement further says that more efforts must be initiated to transfer more transports from trucks to ships.

In Norwegian National Transport plan for 2014–2023, the government will improve sea transport facilities with 19.4 billion kroner (RMB), which is 55 % above the level

Fig. 6.7 Luding bridge, one of China's oldest existing chains or suspension bridges across the Daduhe river, a second class tributary of Yangtze River with an average flow of 1490 m^3/s. It is located in west of Luding county, Shichuan Province, China built in 1706, Qing Dynasty as an important entrance to Tibet with a dimension (span × width) of 103.7 m × 3 m and a maximum height of 14.5 m above the water surface. The bridge body was composed of 13 man-made wrought iron chains with a total weight of over 21 tons

Fig. 6.8 Locks in the Svir River on the water way between Moscow and St. Petersburg, Russia. The first ship in the spring, early in May, is on its way down the river to the Onega Lake with oil. An icebreaker has the previous day made the first opening way on Onega, Europe's next largest inland lake (nr.18 in the world). All together this water way has 19 locks

from the 2013 budget. When the plan was presented the Fishery and Coastal Minister claimed that every ship in average is comparable to 200 trailers [3] (Fig. 6.14).

The Norwegian Statistical Central Bureau (SSB) [4], amongst others reports that in 2011, 3630 Norwegian ships (10 per day) had been recorded in East Asia alone. The corresponding gross tonnage was 123 million tons. In total Norwegian ships with size above 1 000 tons had 63 800 visits to foreign harbors in 2011. Most visits were going to Europe with 38 559 landings, while Scandinavia and the Baltic Sea had 126 million visiting gross tons.

Another investigation SSB tells that petroleum products represented 89 % of inland transport within wet transport. Transport between Norwegian harbors made out 15 % of sea transport in 2007 (Fig. 6.15).

The engineering profession of building good harbors to secure efficient and safe loading and reloading is several 1000 years old. The Romans were pioneers also in this field. In addition to the harbor facilities in Italy, we can mention that Herodias in 20–10 years BC built the harbor city Caesarea on the coast between Tel Aviv and Haifa. The advanced harbor construction used sinking boxes in concrete that were floated into position and later filled with concrete to act as harbors.

Of the 20 largest container terminals today, we actually find 7 in China, 3 in USA, 5 in other places in Asia, and 3 in

Fig. 6.9 On May 10, 2011, an icebreaker has just made the first waterway of the year on the Onega Lake. Seals on the ice are not unusual. The Onega Lake is a key item on the water way between Moscow and St. Petersburg/The Baltic Sea/The Atlantic Ocean

Fig. 6.10 Brekke locks in the Halden water way are claimed to be the highest in Northern Europe. The locks were planned and built by Engebret Soot in 1852. A river flood destroyed the construction in 1861, with a reconstruction ready in 1872. The present locks with concrete chambers, was finished in 1924. The four chambers and three lifts; the boats can be lift boats with length of 26 m, and width 6 m and depth 1.6 m. The lifting is 26.6 m. The Brekke locks are located about 11 km north of the town of Halden

Europe. The world's highest through-tonnage is in Singapore [5]. Antwerp in Belgium is the largest container harbor in Europe, followed by Rotterdam in Netherland, Gioia Tauto in Italy, and Hamburg and Bremerhaven in Germany.

But, Rotterdam is the busiest harbor in Europe in total, as they are on top both with regard to dry and wet cargo bulk transport [6].

How important sea transport is for a country like Norway can clearly be seen from the illustration below. Norwegian export tonnage is about five times as high as the import tonnage, and the sea transport is even more dominant. We have chosen to show the import tonnage as here the many folds in goods' type is highest (Fig. 6.16).

Fig. 6.11 Picture of the old lock system at Brekke, taken during the construction work in 1924. On the back of the picture is written (translated from Norwegian): This picture is taken from the dam and upwards in the old locks. You can see the old upper Brække Dam and the stair with the locks. This is something you will never see again. The water is now 14 m over the old dam with the lock. (The brothers Oscar and Hans Oswald Jahren had both been working at the lock constructions. Hans Oswald had gone to USA, and his brother Oscar sent him the picture with the description)

Norway has a comparatively very long and rugged coast line. In particularly special for the country compared to other countries is therefore the ferry traffic along the coast. A statistic from the Norwegian Statistical Central Bureau for 1999 [8] tells that there were 199 inland ferry lines in operation and that annually transported 46 million passengers. On the inland lakes 12 ferries accommodated 155 000 passengers.

The number of passengers has been fairly stable since the beginning of the 1980s while it back in 1963 only transported 14 million passengers. Some ferry stretches have over years been changed to bridges, and Norwegian bridge structures have been a Norwegian trade mark and have given Norwegian engineers assignments many places in the world (Fig. 6.17).

Ferry traffic is in many ways the opposite of sea transport. For the sea transport we use water as a practical transport lane on rivers, lakes, or oceans, while a ferry most often is used to cross a water hindrance. However, this has from more than thousand years been an important tool to keep the country together. The Chairman of the County Møre og Romsadal, Ingvald Sverdrup, has said about his county:

> In this county about a third of the population lives on island, the rest lives a ferry trip on the other side of the fiord [9].

The shortest ferry connection in Norway is across the Drammen fiord, where it is at its smallest, from Svelvik and Verket. This is only 600 m (Fig. 6.18).

The longest national ferry line in Norway is between the city of Bodø in northern Norway, and the islands of Værøy and Røst with a distance of 193 km. For most of the ferries, we talk about car ferries. There is a big difference between the efficient car ferries we have today and the first car ferry which came into use in 1921 and was a rebuilt fishing boat that got the name: *M/F Bilferjen (Car ferry)*. The ferry was in operation on the west coast near the city of Ålesund, and had space for 3–4 cars [9].

The largest national ferries in Norway are the ferry connection across the Oslo Fiord from the town of Moss to the town of Horten. These ferries in 2010 freighted more than 1.5 million vehicles and about 3 million passengers. They might take 250 cars on each trip, as early as in books from 1582, ferry activity in the same place is mentioned [10], and in a letter from 1712 King Christian IV ordered a new ferry to be built with a capacity to take 16 horses and 50 men. The ferry line is called the Bastøy ferry (Fig. 6.19).

The community of Nesodden with 18 000 inhabitants is located on a peninsula in the middle of the Oslo Fiord, just outside the Norwegian capital, Oslo. The community brags that the people that live there are in a rural area, and still need to take a very short travel to Oslo. Many inhabitants work in Oslo, too. There is a busy ferry connection between to the city with departures several times an hour except during night time that the ferry departs every hour each way. The ferry is a key to giving the population in the community the lifestyle they have chosen. The boat trip from the community to the centre of Oslo takes less than half an hour, and most of the ferries have royal names like *Kongen (the King), Dronningen (the Queen), Prinsen (the Prince), Prinsessen (the Princess), or Baronessen (the Baroness)* (Figs. 6.20 and 6.21).

Not only along the coast have the ferries been an important connection tool, Mjøsa, the largest lake in Norway, got its ferry connection from the east to the west side in 1920 after nearly 20 years of discussion. In 1986, the ferry traffic was relieved by a bridge (Fig. 6.22).

Fig. 6.12 It is idyllic along the Telemark canal even in winter. The picture is from upper Ulefoss

Fig. 6.13 Vrangfoss locks and Ulefoss locks. Repair and maintenance of the locks must take place in the winter months

Fig. 6.14 Ferries are of a wide expression. There are several divisions in difference when the ferry from Oslo, Norway to Kiel in Germany is on its way out the Oslo Fiord, and meets a local ferry on its way in the harbor in Oslo centre

6 Transport Lanes

Fig. 6.15 The sea or oceans as transport lanes makes possible transport in a completely different range of tonnage than on the roads or on rail. However, this necessitates structures in harbor facilities and storage facilities to optimize the transport. Example, grain silos in the community of Buvika, Sør Trøndelag County, Norway

Ferries and bridges are, however, not the only way to cross the water hindrances. In modern time we have the tunnels, but far back to the Viking age some thousand years ago we know, for example, how the ice was used as the transport road. The King of Oppland County in Norway, Halvdan Svarte drowned when he tried to cross the spring ice on Lake Randsfjorden riding home from a party. We also know that the Vikings used winter conditions on the Russian water ways on their journey to Miklagard, the Nordic name for Istanbul some thousand years ago (Fig. 6.23).

We have also had a railway ferry in Norway. The ferry was part of the railway line to the town of Rjukan over Lake Tinn from Tinnoset. The ferry came into operation as early as in 1911, and the ferry was made immortal through the movie about the heavy water sabotage. The Ferry *Hydro*, with heavy water on board to be used for production of atomic bombs in Germany, was sunk during sabotage action during the Second World War in 1944.

The 30-km-long ferry journey along the 460 m deep Lake Tinn was part of an industrial historic railway line for Norway, and to some extent also internationally.

The Rjukan railway line was indeed a complicated transport project with barge transport from the coast at the town of Skien to the town of Notodden (54 km), then a railway line from Notodden to Tinnoset (30 km), and then the railway ferry along Lake Tinn, and with steep mountains dropping into the lake on both sides from Tinnoset to Mæl (30 km), and in the end the railway to Rjukan (16 km).

The rail system got its concession in July 1907, and was opened by King Håkon VII on August 9, 1909. The opening of the line that freighted limestone from the coast to the factory at Rjukan and fertilizers back was one of the foundations for the new city that was established at Rjukan, and the foundation for the largest Norwegian Industrial company, Norwegian Hydro (one of the largest fertilizer companies in the world). The last train on this railway line was in 1991 [11].

Of the seven ferries that have been in traffic on Lake Tin, the last three were: the steam ferry *Hydro*—173 ft—built in 1914 and sunk in the heavy water sabotage in 1944, the steam ferry *Ammonia*—231 ft—built in 1929 and the diesel ferry *Storegut*—287 ft—built in 1956 [12] (Fig. 6.24).

To call the Norwegian coastal steamer (Hurtigruten) a ferry might be wrong. The coastal steamer is extremely important, and in the eyes of many people it is some kind of large cousin to the ferries that are binding both countries together. The coastal steamer has been and is even today the

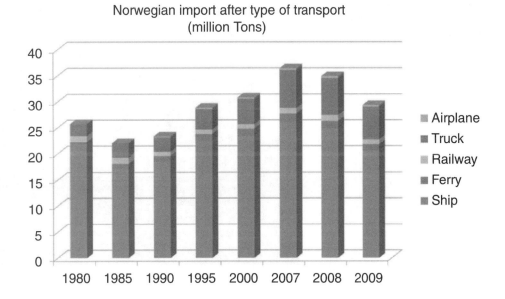

Fig. 6.16 Norwegian import tonnage according to type of transport [7]

Fig. 6.17 The ferry from Norway's fourth largest city Stavanger to Tau takes about 40 min and is for many the start of the coastal road on the west coast from Stavanger to Trondheim. The ferry cannot be regarded as a maritime beauty, but is very efficient

Fig. 6.18 The ferry between Verket and Svelvik is Norway's shortest. The ferry is on its way to Verket. Departure is normally every half hour. The capacity is 18 cars and 90 passengers

Fig. 6.19 The Bastø-ferry on its way in and another ferry on its way out—at the harbor in Moss

Fig. 6.20 The King and the Queen at the harbor in central Oslo, a cold winter day

Fig. 6.21 A ferry leaves Nesodden on its way to Oslo

most important transport lane of the country of Norway. Today, it is the opinion of many, i.e., the coastal steamer is possibly the most important tourist attraction of the country. The introduction of the large coastal steamer companies started in the 1850s, but it took some time before an all year regular route along the coast was established (Figs. 6.25 and 6.26).

The Coastal department tells that Norway has five harbors with highest priority (Oslo, Kristiansand, Stavanger, Bergen and Tromsø) and 32 s priority harbors. About 600 fishery harbors are also organized through the Coastal department [13] (Figs. 6.27 and 6.28).

Looking at the various loading types in Norwegian harbors, we find the following numbers [14]:

- Wet bulk 49 %
- Dry bulk 36 %
- Other cargo 7 %
- Self-driven ro-ro units 4 %
- Containers 3 %
- Nonself-driven ro-ro units 1 %

The list clearly shows the important impact of sea transport on the Norwegian activity in oil, gas, and minerals.

Deep-protected fiords combined with good access to hydroelectric power have been the foundation for the development of a number of important local societies along the long coast of Norway. In addition to the raw material bases and the energy possibilities from water power, efficient

Fig. 6.22 Ferry—meeting in the Geiranger Fiord. The car ferries from Geiranger and Hellesylt are meeting. This is also an interesting object for photography for interested tourists. The mountain walls are going so steeply into the deep fiord that the road building is impossible, so the ferry is the only connection along some of the fiords

Fig. 6.23 It is in early April. The ice has started to melt on Norway's fourth largest lakes—Randsfjorden. The transport on the ice has to stop

sea transport has been the key competitive factor in an international competition. One of the most typical examples for these local industrial establishments is the town of Odda in the South Fiord in Hardanger on the south coast. This town had three important industrial plants based on various local resources (Fig. 6.29).

Norway has always had an important position as a sea transport nation. The picture of the largest sea transport nation is a bit different depending on the statistic of where the ships are recorded and the statistic over where the shipping companies come from. The picture also changes somewhat over time. The shown statistic is from 2007, but gives a realistic general advice (Figs. 6.30 and 6.31).

If we look at the values of the world merchant fleet, the picture changes a bit (Fig. 6.32). The newspaper Dagens Næringsliv on January 30, 2013 gives the following numbers:

Probably the most busy transport road in the world is the **Yangtze River**, running into a delta near the world's most busy transport harbor, Shanghai. Shanghai had in 2010, 23 million inhabitants, while the city had 13 million. Shanghai is the world's third biggest container harbor, but has from 2005 been the biggest harbor measured in total tonnage—

Fig. 6.24 At the end of the old railway track at Mæl, north in Lake Tinn we can still find the protected railway ferry "Storegut". At the pier beside is the ferry "Ammonia". Both D/F Ammonia and M/F Storegut were declared protected by the Norwegian government in 2009

537 million tons in 2006 [17]. The name of Shanghai also is closely connected to the sea. The reason why Shanghai has such a gigantic tonnage is based on the fact that the Yangtze delta generates 20 % of the gross national product in China. Along the Yangtze River we also find one-third of the population in China, and more people than in Europe all together (Fig. 6.33).

Yangtze has river transport all the way up to Yibin, boats up to 10 000 dwt. can come up to Nanjing, boats up to 5 000 dwt. up to Wuhan, 3 000 dwt. to Yichang and 1 500 dwt., to Chongqing (see also Chap. 4.)

One of the side rivers to Yangtze is Xiangjiang. During the Qin Dynasty for over 2000 years ago, the local population constructed a canal between Xiangjiang and Lijiang where these rivers were closest to each other. Lijiang runs into the third largest river in China, the Pearl River. The Pearl River runs into the sea through the Pearl Delta, near the large cities like Guangzhou, Shenzhen, Hong Kong, and Macao. Part of this canal that binds these two rivers together is still in use.

Yellow River acting as water conservation and inland river navigation can be traced back to 3000 years ago. Records on the earliest long distance water transport along Yellow River appeared in late West Han period (206 B.C.–25 A.C.) and reached its summit in Tang Dynasty (618–907). This has been continued later for more than thousand years. However, Yellow River as the second largest river of China functions much less than Yangtze River as inland waterways navigation. The reason might be

- Low water flow and twisty river course;
- More silt in the water and less renovation of the channel;
- Less developed in economy;
- Frequent cutoff of the waterways due to excessive exploitation and utilization;
- Different weather conditions—as 4 months of frozen period for Yellow River.

The people along the Yellow River, however, have been long struggling in using the waterway as transport lane to suit local conditions. From the origin of the river in the Bayan Har Mountains in Qinghai Province of western China to Liujia Gorge in Yongjing County of Gansu Province, the river winds through numerous mountains and valleys for 1200 km with a total fall up to 1300 m. During the Republican period of early twentieth century of China, people in the upper stream of Yellow River tried very hard to use cow skin or lambskin rafts and wooden rafts to carry grains, oil, etc., downwards (Fig. 6.34).

A folk song of Shanbei (North of Shaanxi)—*Song of Yellow River Boatmen*—is very popular in Shaanxi Province along the Yellow River, which goes as

Fig. 6.25 MS Lofoten—built in Oslo in 1964 was in 2013 the oldest of the coastal steamer fleet (length 87.4 m, width 13.2 m, for 340 passengers with 153 sleeping beds). In the summer it is a tourist attraction and mostly fully booked. A trip into the Geiranger Fiord increases the journey attraction further. Here Lofoten has just left its passengers, and is on its way out from Geiranger. An important function for the coastal steamer is mail transport, and then the mail transport flag has to be carried in the back

Do you know how many dozens of bends the Yellow River of the world has? How many boats are there in these dozens of bends? How many poles are there in these dozens of ships? How many boatmen are there to ride on boats?

I know there are ninety nine bends in Yellow River of the world, ninety nine boats on the ninety nine bends, ninety nine poles on the ninety nine boats, ninety nine boatmen ride the boats.

Follow the piano accompaniment of the Song of Yellow River Boatmen, have not you imagined through this simple and deep, terse and forceful song pouring down of the twisty Yellow River, the portrait of the boatmen struggling in riding the boats through the waves?

Water transportation becomes very active after the Yellow River enters Shandong Province from Henan Province where water has become rich and economy is more

Fig. 6.26 In Norway you have to buy a ticket for the ferry, but in the neighboring country Sweden, that also have some ferries, you do not pay in the ferries on the main roads. The Picture is from the Sund-Jaren ferry across the Lake Stora Le—a ferry distance of about 10 min, in the county of Dalsland, Sweden only 10 km from the Norwegian border

Fig. 6.27 The harbor of the Norwegian capital, Oslo is not the busiest with respect to cargo transport, but is a busy harbor with respect to ferry transport and private boats—and the harbor is beautiful

developed. The sailing boat can travel in a day more than 100 km with tail wind and 20–30 km against the wind (Figs. 6.35 and 6.36).

In 2009 Report on Yellow River Navigation Programming was issued by the Ministry of Transport calling on a comprehensive and unified act for the nine provinces and autonomous regions on the development and construction of the Yellow River navigation. The target by 2020 is to achieve sectional navigation of 300-ton ships and by 2030 the access to the sea navigation for 600-ton ships.

China's Grand Canal, the longest artificial waterway in the world with a history of more than 2 400 years, was approved in June 2014 by UNESCO on the World Heritage list. The 1,794-km-long canal, which is nine times longer than the Suez Canal runs from Beijing to Hangzhou, capital city of China's eastern Zhejiang province and covers eight Chinese provinces and municipalities and 15 % of China's population living in areas along the canal. It is also the second most important inland water transport, after the Yangtze in China.

The Grand Canal in history was first constructed in sections starting from Hangzhou in 486 BC during the period of Spring and Autumn (770–477 BC) and continued in Warriors Period (475–221 BC). The second phase of this huge

Fig. 6.28 The harbor in Bergen, Norway—seen from a local hilltop, Fløyen, a rainy day

Fig. 6.29 Odda has a fantastic location in the bottom of the South Ford in Hardanger, and right under the glacier of Folgefonna

project happened in Sui Dynasty (581–618) with Luoyang as a start to further build a multibranch canal. Later in Yuan (1271–1368), Ming (1368–1644), and Qing Dynasties (1644–1911) the project continued and formed the whole structure of the huge canal. This is not only connecting China's five river systems, including the Yangtze River, Yellow River, Haihe, Huaihe, and Qiantang Rivers, but also witnessing the country's cultural communication and mixing of different ethnic groups. It therefore served in history as an important carrier of materials and ideas between the capital cities and economic cities. No wonder the canal is the world's largest and most extensive civil engineering project prior to the Industrial Revolution (Figs. 6.37 and 6.38).

Of the transport capacity in China in December 2004, 1.8 billion tons among the total of 16.1 billion tons took place on the water ways [33]. This was an increase of 18 % from the year before, nearly double of the increase in cargo transport in total. Another statistical data from China from Water

Fig. 6.30 The merchant fleet in the biggest sea transport nations in million tons according to registration [15]

World Merchant Fleet–acc. to. Registration mill tonns

COUNTRY/REGION	Million tons
Panama	141,8
Liberia	59,6
Bahamas	38,4
Singapore	31,0
Greece	30,7
Hong Kong, China	29,8
Malta	23,0
China	22,3
Cypros	19,0
Norway	17,5

Fig. 6.31 The merchant fleet in the largest sea transport nations in million tons according to ownership [15]

COUNTRY/REGION	Million tons
Greece	95.4
Japan	89.3
Germany	54.4
CHINA	41.5
USA	36.0
NORWAY	33.4
Hong Kong, China	36.4
UK	22.1
South Korea	19.3
Taiwan, China	16.1

Transport on December 23, 2009 [21] gives the percentage of freight turnover of inland river transport among the total transport volume of highway, railway, and inland rivers as 7.83, 6.06, and 8.19 % for 1985, 2000, and 2005, respectively. This indicates an increase in tendency even under the remarkable improvement of other two transport systems. These figures also give clear signals that water transport is far from a dying industry. In addition, we need to know that the investment cost of inland transport is only one-third of the railway and 1/7 of the highway investment. The specific transport volume per horsepower of the inland transport is 4 times higher than railway and 50 times higher than highway, while the cost of inland river transport is only 1/2 of railway and 1/3 of the highway. The China's inland river transport

Fig. 6.32 The value of the world merchant fleet per 31.12.2010 [16]

World Merchant Fleet –Fleet values million USD

COUNTRY/REGION	Million USD
Japan	119 925
Germany	95 330
Greece	81 239
Norway	53 758
USA	51 968
China	50 892
Denmark	33 277
Singapore	27 477
South Korea	23 623
Hong Kong, China	19 903

Fig. 6.33 Barges and boats are on their way up the Huangpu River in Shanghai. It is high tide and the traffic is on their way up the river. At low tide the traffic goes down the river. Huangpu is 114 km long, and is part of the Yangtze delta

therefore plays an important and irreplaceable role and is very competitive in many aspects. It has a big potential in the future, especially for Yellow River transport.

It is not like this on all the water ways in the world. Historic water ways sometimes lose in the competition to new structures as for example railway transport.

In an article in Concrete International in December 1996 [22] Luke M. and Billie G. Snell tell a fascinating story about one of the first really big concrete works in USA.

One of the most fascinating canals and water ways in history is the **Erie Canal.** The canal was built by the USA Senate in 2000 recognized as

> the most successful and influential human-built waterway and the most important works of civil engineering and construction in North America [23].

During the beginning of 1800, the transport problem across USA was considerable, and the cost to cross the Appalachia Mountains was a great economic burden.

Fig. 6.34 People living along the upper stream of Yellow River during the Republican period of early twentieth century of China used combination of horse for land transportation and log raft for water [18]

Fig. 6.35 Song of Yellow River Boatmen, Piano Accompaniment

Fig. 6.36 Boatmen on Yellow River, China [19]. Courtesy of Li Sun and a commemoration for Mr. Sun who passed away on March 28, 2016 devoted his life to the beloved homeland of Shaanxi

Fig. 6.37 Wuzhen section, part of the Beijing-Hangzhou Grand Canal, located in Wuzhen, or Wu Town in Zhejiang province, a famous China ancient scenic water town[20]. The Grand Canal is a man-made inland waterway that runs through north and south in eastern China linking 5 water systems. It is the longest man-made waterway in the world stretching over 1797 km and passing through 4 provinces and 2 municipalities.

[20] http://news.cnnb.com.cn/system/2014/06/21/008093282_07.shtml

A wagon train with eight horses needed to take 15–45 days on the trip from Buffalo to Albany with a cost of 100 USD per ton [22] (Fig. 6.39).

It was therefore a number of reasons for the official representatives of the New York State to be interested in a Canal Project from the Hudson River near Albany to Buffalo at the Lake Erie, the Erie Canal. When the idea was put forward to President Jefferson, however, it was blankly turned down and called pure madness. However, on July 4, 1812 the State of New York awarded 20 000 USD for the project for a detailed study.

The canal from Albany to Buffalo was 584 km long and had 36 locks when it was opened on October 26, 1825.

There are many stories and anecdotes about the construction of the canal. One of them is about the self-educated engineer Canvas White who in 1816 was employed as an assistant. Later he became a key person in the building of the canal. The Governor of New York was De Witt Clinton, and he demanded that the study trip should be made to England to study existing canals and locks and to buy the most modern surveying equipment that could be found. White went on this trip, and he used his time on other side of the Atlantic Ocean well with a number of detailed studies.

It was decided that the locks should be built by natural stone with lime mortar in the joints, with exemption of the most difficult parts. Therefore, hydraulic cement should be used. Hydraulic cement had to be taken from Europe in the beginning, and was very costly. White had known cement production under his study trip to England. He found acceptable limestone near Chittenango, southeast of Syracuse that could be used to produce cement in a safe manner. This "White-cement" was used in the canal construction already in 1818, (The patent on Portland cement was handed in by Joseph Aspdin in Leeds, England in 1824) and the production kept on to 1840. Even if White patented his simple cement, he did not trust that he had a superior technology. Cement from Europe was used in the most important parts of the canal.

Fig. 6.38 Southern part of the important communication channel connecting northern and southern China in ancient times still functions well today and is considered a special "alive" heritage [20]

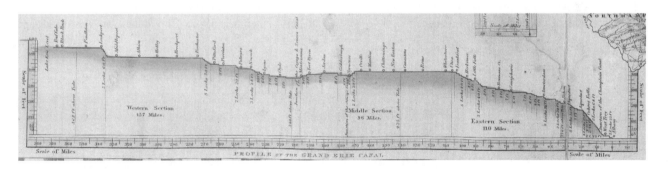

Fig. 6.39 Profile of the original Erie Canal

Between the years 1834 and 1862 the canal was extended, and in 1918 the canal was extended again, and partly modified to what is called the New York State Barge Canal. The original canal was 12 m wide and 1.2 m deep. The soil that was dug out was deposited on the side of the canal to form a towing road so that horses could tow the barges. It is claimed that during the construction this was a swampy area, where 1000 men died in one season due to the mosquitos and swamp fever [6]. During the extension of the canal that started in 1834, it was extended to a width of 21 m and depth of 2.1 m. The New York Barge Canal that was completed in 1918 extended the canal to a width of 37 m and depth of 3.7 m. Today the canal includes a canal system of 843 km of water way.

During the great emigration period to America in the middle of the 1800s, the Erie Canal was a key nerve in the transport from east to west. While there in 1855 were registered 33 000 shipments on the canal, the number of commercial shipments was down to 42 in 2008. However, the canal is still popular for pleasure boats.

The Erie Canal has all together 36 locks to handle a height difference of 169 m.

The America Boats was an important expression for many Norwegians in several generations, and many had a relationship with them often because one or more in most families had been on board with one of them in one relationship or the other.

The first "modern" Norwegian emigrant to America crossed the Atlantic Ocean in 1636 and 1637 with Dutch ships, and they used, respectively, 2 months and half a year on the trip.

However, the 52 persons who went across with the sail ship *Restaurationen* in 1825 marked the real start of the Norway–America emigration.

The emigration culminated in 1882 with 29 000 Norwegian emigrants to America in 1 year.

The dream about America was great among people in the poor country as Norway, at least for over hundred years. The result is that there are more people with Norwegian origin in America than in Norway.

However, most people connect the name *The America Boats* with the twentieth century and the eight ships that sailed the flag of The Norwegian America Line. This shipping company was found in 1910 and the first ships came in 1913. The shipping company was in more than one way a symbol of national pride and independence.

The ships were an important tool on the first part of the journey for emigrants, and also a transport media for those that had enough money to take a trip across the Atlantic Ocean for pleasure, or to visit relatives. In addition to the America trade, the ships were also used to cruises like journeys along the Norwegian coast in the Baltic Sea, and later to further destinations. The ships provided also an attractive job possibility. Some of the men that got jobs also jumped off when the boat came to America. One of them was my father that went ashore from the ship *Bergensfjord* in New York in July 1924, and did not return to Norway until 1934. *Bergensfjord* was also called the lucky ship because it survived two world wars. *Stavangerfjord* was the ship that sailed the longest time in total 45 years. The last trip was the Christmas trip in 1963, the 70s crossing of the Atlantic Ocean.

When *Oslofjord* came back from New York on the "virgin-voyage" as "Christmas-boat" in 1949, she had set a new speed record across the Atlantic from Ambrose to Marstein lighthouse with 6 days, 10 h and 31 min. It was school-free and party in the harbor, and we with admiring eyes looked at the wonder [24] (Fig. 6.40).

The emigration from Norway to America was reduced in the 1920s and 1930s. The top year for the America ships was in 1923 with 18 000 west bound passengers. Then the American Congress decided an immigration law that reduced the Norwegian quota considerably. To compensate for the loss of passengers, the NAL ships started to also port in Halifax in Canada. In the top year 1927 they had 3212 passengers for this destination. In the period from 1926 to 1930 the emigration from Norway to Canada was 13 492 persons, of these 10 416 were transported on the America ships [24]

Look to Norway, President Franklin D. Roosevelt of USA said in September 1942. The quotation was used and

Fig. 6.40 The America ships [24]

NAL's America ships

NAME	Birth	Tonnage	End
S/S Kristianiafjord	1913	10 625 BRT	1917.-Wrecked
S/S Bergensfjord (1)	1913	11 015 BRT	1946 – Sold, demolished 1959
S/S Stavangerfjord	1918	12 762 BRT	1964 – Soldand demolished
M/S Oslofjord (1)	1928	18 765 BRT	1940 – Mine blown up
M/S Oslofjord (2)	1949	16 844 BRT	1969 – Charteredout ,Fire 1970
M/S Bergensfjord (2)	1956	18 739 BRT	1971 – Sold, 1980 - Fire
M/S Sagafjord	1965	24 000 BRT	1983 – Sold(("Saga Rose")
M/S Vistafjord	1973	24 292 BRT	1983 – Sold ("Saga Ruby")

Fig. 6.41 Cruise ship at anchor, Villefrance, France

misused many times, and was first and foremost used to describe the work of Norwegians on the world oceans. The contribution and effect both international and for national economic and cultural development as a sea nation were formidable.

Before the Second World War, the very small country of Norway had built up one of the biggest merchant fleets in the world, which not only had activity on all oceans, but also had activities and specialties one a number of transport areas. On the oceans Norway had become one of the most important nations in the world, and could both with respect to number of ships and in tonnage to be compared with the largest shipping nations. If looking at the numbers in a per capita context, it was much larger than any other nation. However, of the Norwegian merchant fleet 694 ships, 47 % in number of ships or 34 % of the tonnage got lost in the war from September 1939 to May 1945 [25]. After the Second World War there were few Norwegians who did not had family members or other types of relatives that had experienced one ship wreck or more. A cousin of my mother came home in 1945 after 6 years away from home with six ship wrecks in his luggage. His old typewriter is still a souvenir in my office. Heavy and heartbreaking happenings today get another dimension in the memory of what happened at sea in those years.

The oceans, the sea, or to sail along the coast have been an activity that has its place in the spine of most Norwegians. Probably this sea syndrome has been somewhat reduced over the last tens of year. In the first 2/3 of the last century it was difficult to find people that by themselves or through family members or friends had not experienced at least a few months at sea.

It is difficult to say what part of the Norwegian ocean activity has been most important or impressive. However, at least one point is in many folds and creative and knowledge-based ability find new special field to utilize the oceans in a transport possibility (Fig. 6.41).

- Tankers account for the largest part of the Norwegian tonnage, with specialties as gas tankers and ships for chemicals.
- All types of bulk ships including ore ships and car freighters.
- *Various* types of dray loading ships including cooling- and freezing ships, container ships and ro-ro ships.
- Offshore ships with specialties as service ships and supply ships.
- Passenger ships, pure cruise ships, and ferries.
- Whaling ships were also in many years a Norwegian specialty.

What originally started as the needs for the fishing and farming villages to freight their good to the cities and the export harbors, and to bring the city goods back has developed to one of the most important industries for the Norwegian society.

In addition to the many seaman schools, many Norwegian sailors also got their first experience at sea through the school ships—the old sail ships: Christian Radic, Sørlandet, and Statsraad Lehnkul (Figs. 6.42 and 6.43).

Fig. 6.42 The sail school ship of the city of Bergen, "Statsraad Lehmkul" at anchors in its home town. The bark was built in Bremerhaven in Germany in 1914, and sailed in many years as school ship in the German merchant navy. During the First World War it was taken prisoner by Great Britain. In 1921 the bark was bought by the previous member of the Norwegian Government Statsråd Kristoffer Lehmkul

Fig. 6.43 A lot of Norwegian shipping history is anchored up at Bygdøy, a rural peninsula in the capital Oslo. Here we find the Viking ship museum, and near the sea we find Norwegian Maritime Museum, together with Kon Tiki, and the polar ships Fram and Gjøa

As a contra point to the many specialties in sea transport we would like to mention two fascinating examples of series production of ships when this is needed.

The Liberty Ships

Both during the First and the Second World Wars, torpedoes sank a number of ships, which led to lack of transport tonnage on the oceans. New boats had to be built and faster and cheaper than during peace time. The British government ordered, in 1940, 60 new ships from USA to compensate for the ships that had sunk during the first phase of the war. The ships should be of a size of 10 000 tons, and originally of an English design based on coal firing because coal was an English energy source. This construction was modified by the USA's Marine Commission, amongst others to use oil as energy source, and simplified to get them cheaper and faster in production. Welding was also taken into use while the original English concept was to use riveting. A small, special but important part of the construction was the great number of female welders used under the construction work because the ordinary wharf workers were called in for war service.

Fig. 6.44 This table was bought in Long Island, USA in 1972. The table top shows that it is an old ship hatch. As such it has probably lived an interesting, journey rich and hard life. This table, previously a cargo hatch from a Liberty Boat now lives a quiet life in a summer house on an island in the Oslo Fiord. *Photo* Torild Jahren Petersen

The American version got the name *EC2-S-C1* (EC for Emergency Cargo, 2 for ship between 120 and 140 m length, S for Steam ship, and C1 as design number).

In total, there were 2 710 *Liberty* ships built at 18 American wharfs in the years 1941–1945. The construction time was impressive. The first boat took 230 days to build. Later, the average time from stretching of the keel to launching was cut down to 42 days. The construction record was made by *SS Robert E. Peary*, that was built in 4 days and 15.5 h [26].

In the beginning, the Liberty boats did not get any good reputation. President Franklin D. Roosevelt called them *a dreadful looking object*, and *Time Magazine* called them *Ugly Ducklings*. The first boat was launched on September 27, 1941, and to improve the image of the boat, the day was named the *Liberty Fleet Day*. President Roosevelt quoted a speech from Patric Henry from 1775 and said *Give me liberty or give me death*. That is why the name of the boats.

A number of the boats got material problems during the war, and many believed that was due to the introduction of welding. An investigation commission, however, concluded that the weakness came from fatigue in the material in some of the steel deliveries that was not able to handle the low temperature combined with the vibrations.

Many ships were lost, 2 400 Liberty ships survived the war. After the Second World War, many ships were sold to different countries. Greek shipowners bought 526 boats and Italian 98 boats, something that made a foundation for the flowering of several large and well-known shipping companies.

Most of the ships were after sometime cut up around the world, and not only the steel were recirculated. A company at Long Island bought loading hatches in large numbers and used the solid, beautiful parts as tabletops for writing tables and coffee tables (Fig. 6.44).

The Concrete Ships

Both the First and the Second World Wars led to lack of raw materials and lack of steel in particular. The Norwegian Nic. Fougner in 1917 built the first ocean going concrete ship: Namsenfjord, 400 tons from a wharf in Moss in the Oslo Fiord in Norway. The first plan was dated in 1912, but lack of financing led to that the plan had to be postponed (Fig. 6.45).

Before the construction of *Namsenfjord*, Fougner had amongst others built concrete barges in Manila in the Philippines and later he designed and built also other barges, tow boats, and lighthouse ships.

Fougner's initiative led to construction of concrete boats in a number of countries in the years afterwards, and in 1922 Fougner published a book on the topic; *Seagoing and other Concrete Ships* [27], that tells about the development.

Due to the shortage of steel during the First and Second World Wars, a number of concrete ships were produced in series. In USA they built in the period from 1918 to 1919, 12 concrete ships from 2460 to 6380 tons, whereas seven oil tankers of 6380 tons (Fig. 6.46).

During the Second World War 104 concrete ships were built in USA with capacities from 3 200 to 140 250 tons. Of these, 24 were larger ocean going vessels. This is comparable to 45 Liberty ships [27]. In Great Britain they at the

Fig. 6.45 Copy of original drawing of Namsenfjord [27]

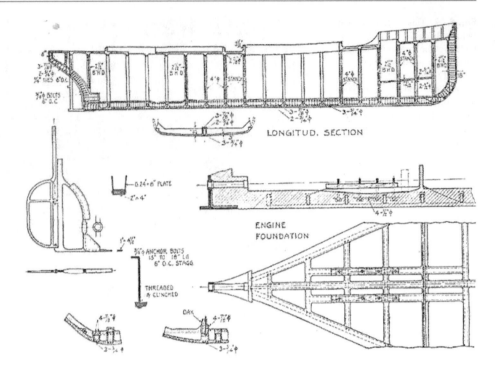

Fig. 6.46 The oil tanker U.S.S. Selma, launched sideways in Mobile, Alabama in June 1919

same time built more than 100 concrete ships in series. The boats were built in elements and were called "Crete" boats. Three of these ships later stranded in Norway.

From 1942 to 1943 it was along the English south coast and up along the rivers that constructed and in series built in secret a great number of concrete barges of two variations. On the D-day of June 6, 1944, these barges were floated across the Channel to France and became the foundation to the landing bridges and the protection breakwater for the *Mulberry Harbor,* where the landing in Normandy in France took place. At the same time and also in deep secret were 64 m long and 10 m wide oil barges built in Japan, which got the name *Takechi Maru*. The oil barges went in oil trade to Sumatra. The barges still existed a few years ago, now as wave breakers in the fishery harbor of Kure on Hiroshima [28, 29].

In 1970 there was a surge in China making cement ships due to lack of iron and steel. As said a world largest (actually it was not) load of 3000-ton cement ship Gutian was built in 1974 with a dimension (length × width × depth) of 105.2 m × 14.5 m × 8.1 m and a waterline of 5.7 m and a displacement of 5773 tons. The ship was designed for all

Fig. 6.47 Picture of the news report in 1973 on celebrating Gutian cement ship's successful maiden voyage [30]

Fig. 6.48 Pictures from Xinjing News of China reporting the demolishing of Gutian cement ship which ended its final destiny on January 16, 2013 [31]

types of packaging cargo and bulk cargo transport. The ship, after finishing its maiden voyage, was abandoned for long-term voyage due to its too big self-weight and high fuel consumption. It was finally demolished in early 2013 after nearly 40 years' abandonment (Figs. 6.47 and 6.48).

It might belong to the history that about 10 % of the raw material to the concrete in the concrete ships is water, which is needed for cement to hydrate. Today about 2 billion tons are needed to produce concrete every year.

An important part of the sea transport history is the lighthouses that guide the sea travelers in fog and darkness. The Pharos Lighthouse at the inlet to Alexandria in Egypt was built nearly 300 BC and is regarded to be the first in the world. It is also regarded to be one of the seven wonders of the world. Through history many interesting lighthouses are told and their importance for ship transport is considerable. In the last 25 years of the sail ship era up to about 1880, nearly 50 000 British ships were shipwrecked, and as late as

Fig. 6.49 Norway's southernmost point on the mainland—Lindesnes. The lighthouse can barely be seen in the fog. The purpose of the lighthouse was to give the ships guidance at night and foggy weather

in 1881 nearly 1000 ships went down. From then on the conditions improved tremendously, thanks to two relationships happening in the same period: the transition to steam ships and the building out of the lighthouse system.

The lighthouse system in Norway started in 1656 when a big fire was lit on the southernmost point in country, Lindesnes, to guide ships on their way into Norway (Fig. 6.49).

Fig. 6.50 Has not this drawing aroused the memory of innocent childhood of yours? Drawing exercise: Yi Gong

Now let us relax and entertain ourselves with a poem about boating written by Yang Wanli (1129–1206), one of the top four poets in Song Dynasty featuring on pastoral poetry. This poem uses even though so simple words, it yet showcases a vivid, interesting picture as below full of wisdom and childlike fun, through which we also feel the author's affection for children and careful observation to life (Fig. 6.50).

Boating through Anren
by Yang Wanli

In a fishing boat there are two innocent kids.

They sit in the boat withdrawing the pole and stopping the oars.

No wonder they open umbrella while it does not rain.

It is nothing with rain but to use the wind to drive the boat forward.

References

1. Internet 12.12.12: *Grand Canal (Ireland)* http://en.wikipedia.org/wiki/Grand_Canal_(Ireland).
2. Internet 26.09.12: *Nasjonal transportplan, Sjøtransport, Fiskeri – og kystdepartementet, (National transport plan, sea transport, The Fishery – and Coastal Department)* http://blogg.regjeringen.no/ntpdiskusjon1012/fiskerihavner-og-og-farleder/.
3. Hans Christian Færden: *Sjøveien skal få motorveistandard. (The sea transport shall get motorway standard)* Ingeniør-Nytt, Oslo, Norway Nr. 05– 13.
4. Internet 26.09.12: Statistisk Sentralbyrå (Statistical Central Bureau): *123 millioner bruttotonn i Øst-Asia. (123 million gross tons in East-Asia)* http://www.ssb.no/skipanut/main.html.
5. Thoresen Carl A.: *Port designer's handbook*, Thomas Telford Ltd., London 2010.
6. Internet 27.09.12: *Maritime transport statistics.* http://epp.eurostat.ec.europa.eu/statistics_explained/index.php/Maritime_transport-sta.
7. Statistisk Sentralbyrå (Central Statistical Bureau): *Statistisk årbok (Statistical Yearbook) 2011,* SSB, Oslo/Kongsvinger, Norway, August 2011.
8. Internet 26.09.12: Statistisk Sentralbyrå: *Regular coastal trade 1999,* http://www.ssb.no/rutinn_en/main.html.
9. Foss Bjørn: *Ferjelandet (The ferry country),* Forlaget Nordvest, Ålesund, Norway, 1986.
10. Internet 01.12.12: *Fergesambandet Moss – Horten (The ferry Connction Moss- Horten),* http://no.wikipedia.org/wiki/Fergesambandet_Moss%E2%80%93Horten.
11. Gary Payton og Trond Lepperød: *Rjukanbanen,* Hefte nr.1. Norsk Industriarbeidermuseum forteller 2009. Rjukan, Norway.
12. Internet 26.09.12, Statistisk Sentralbyrå: *Færre utlendinger reiste med fergene (Fewer foreigners used the ferries).* http://www.ssb.no/ferge/main.html.
13. Internet 29.09.12: Kystverket (Coastal Department): *Havner (Harbors),* http://131.253.14.66/proxy.ashx?h=SqUs0jwpolTUQkgENoBV1UKHP8nS8LbY&a=.
14. Internet 29.09.12: Statistisk Sentralbyrå: *Havnestatistikk (Harbor statistics),* http://www.ssb.no/havn/tab-2012-08-29-01.html.
15. The Economist: *Pocket World in Figures, 2007 Edition,* Profile Books Ltd., London, England.
16. Dagens Næringsliv: *Flåteverdier (Fleet values),* DN 30. Oslo, Norway, January 2013.
17. Internet 15.11.12: Wikipedia. *Shanghai,* http://wikipedia.org/wiki/Shanghai.

18. Access on June 3, 2014. http://blog.sina.com.cn/s/blog_65cefba50100hq21.html.
19. Access on June 3, 2014. http://www.cd-pa.com/bbs/thread-70256-1-1.html.
20. http://news.cnnb.com.cn/system/2014/06/21/008093282_07.shtml
21. Access on June 3, 2014. http://epaper.zgsyb.com/html/2009-12/23/content_6956.htm.
22. Snell Luke M. & Snell Billie G.: The Erie Canal – America's First Concrete Classroom, Concrete International, December 1996.
23. Internet 25.11.12: *Erie Canal,* Wikipedia, http://en.wikipedia.org/wiki/Erie_Canal.
24. Holm Yngvar: *Den store boken om Amerikabåtene. Nasjonens maritime stolthet.* Edvarde'en Forlag, Bergen 2004.
25. Berggren Brit, Christensen Arne Emil, Kolltveit Bård: *Norsk Sjøfart (Norwegian Shipping Industry) – Bind 2,* Dreyers Forlag AS, Oslo, Norway, 1989.
26. Internet1.01.13: *Liberty ship,* http://en.wikipedia.org/wiki/Liberty_ship.
27. Fougner Nic. K.: *Seagoing and other Concrete Ships,* Oxford Technical Publications, London, 1922.
28. Jahren Per: *Concrete- History and Accounts,* Tapir Akademisk Forlag, Trondheim, Norway, 2011.
29. Yohio Kasai, Nagataki Shigeyoshi; *One Hundred people Who Supported the Concrete Technologies of Japan,* Cement Newspaper Co. Ltd, (på japansk) Diverse avsnitt oversatt av Takashi Sumida.
30. Access on June 7, 2014. http://news.163.com/photoview/00AP0001/21657.html.
31. Access on June 7, 2014. http://epaper.bjnews.com.cn/html/2013-01/16/content_403989.htm?div=-1.
32. Sonia Kolesnikov-Jessop, Did Chinese beat out Columbus? China Daily: Saturday, June 25, 2005.
33. Internet 21.01.13: *Pacific Ocean,* http://en.Wikipedia.org/wiki/Pacific_Ocean.

The Energy Source

Water, or rather water in movement, water with movement potential, and water in a transition phase, are renewable energy sources that are not only important, but also with increasing importance nearly all over the world. In some countries, water is even the source for all electricity supply.

The history behind our electricity supply in many ways starts with the water wheel.

The first reference to its use dates back to about 4000 BC, where, in a poem by an early Greek writer, Antipater, it tells about the freedom from the toil of young women who operated small handmills to grind corn. They were used for crop irrigation, grinding grains, supply drinking water to villages and later to drive sawmills, pumps, forge bellows, tilt hammers, trip hammers, and to power textile mills. They were probably the first method of creating mechanical energy that replaced humans and animals [1].

The earliest record on waterwheel in Chinese history comes from the book **Zhuangzi On the Heaven and the Earth**, an ancient Chinese text from the late Warring States period (476–221 BC) which tells that an appliance named "Gao" (i.e., shadoof, or shaduf) was invented by Zigong, one of the 72 disciples of Confucius in the fifth century BC. It is the earliest prototype of waterwheel working like a lever: with a crossbar supported on the bracket, a water barrel hangs under one end while heavy objects hang under the other end. This had been used since then for 1000 years in the rural areas of ancient China.

It is worthy of mentioning that Chinese inventor Zhang Heng (78–139), Eastern Han Dynasty (25–220), used a water wheel in 132 as power and invented the first celestial globe in history to apply motive power in rotating the astronomical instrument of an armillary sphere. Zhang Heng had a bronze celestial globe constructed on the basis of a bamboo model. He improved it by remapping over 2000 stars and powering it so that it would make one rotation per year. He employed a complicated gear system to link the armillary to a kettle clepsydra or water clock. As the water dripped from one pan to another, the weight would drive the gears and the sphere would advance. He added a gear-driven device that demonstrated the waxing and waning of the moon [2].

After Zhang Heng, Ge Heng of the Wu (229–280) in the Three Kingdom Period and then Qian Lezhi (Liu Song Dynasty 420–478) also developed water-powered armillary spheres.

In 721 AD, the monk Yi Xing and a military engineer Liang Lingzan and others designed and made a water-powered armillary sphere that was an improvement on Zhang Heng's. This one could not only demonstrate apparent motion of celestial bodies, but also install two wooden men who hit a drum to sound the time. It used an escapement method to equally divide the power from a water wheel and transmit it to the armillary. The stars could move according to time but the planets also moved in their irregular patterns by means of a series of bronze wheels. They installed it on the grounds of the palace in the capital Chang'an, i.e., today's Xi'an, capital city of Shaanxi Province, China [2].

The peak of the technology (1092 AD) came during the Song Dynasty (960–1127–1276). During the Northern Song Dynasty (960–1127), an amateur astronomer, Zhang Sichun, made a huge water-powered armillary in 979 AD on the basis of the previous armillary spheres designed in the Han and Tang dynasties. It was 3.3 m high. Su Song and Han Gonglian later in 1092 made the most complicated and most elaborate powered armillary sphere [2].

The true meaning of waterwheel was invented in 184 by Bi Lan, an eunuch during the time of Late Eastern Han (184–220). It was named as "Fan Che" and first used for street flushing demand by taking water from river and later for agricultural irrigation. This type of waterwheels can be driven not only by hand turning and foot pedaling but also by cattle, wind, and flowing water itself with the same design in structure as modern waterwheels. In China, Folk Culture Village located in Chexi Tu Nationality village, Yichang, Hubei Province, they exhibit 38 types of waterwheels showcasing the more than 2000 years history of waterwheel in China.

One typical example of waterwheel application happens in Lanzhou, capital city of Gansu Province, northwestern China, where Yellow River passes through the city. Geographically Gansu Province is located in the upper stream of Yeller River where lands along the river banks are 10–80 m higher than the water surface, and thus makes the irrigation difficult. The scarce in rainfall in the region further worsens the agriculture.

A scholar, Duan Xu, during the Jiajing Period (1521–1567) of Ming Dynasty, invented waterwheel as an effective means to enhance irrigation. This was advocated and carried forward for more than 400 years. Based on the statistics of the local government in 1944, there were 361 waterwheels in the range of 350 km along the Yellow River with a total irrigation coverage of 96 280 μm (1 μm = 666.7 m^2). The average coverage of irrigation per waterwheel is 267 μm while the largest one with a diameter of 24 m can reach as high as more than 400 μm [3].

Now waterwheels have become one of the Lanzhou city's icons. Lanzhou Waterwheel Park (Shuiche Yuan), located in the south bank of Yellow River of the city, demonstrates uniquely shaped waterwheels that were invented during the Ming Dynasty. The entire area used to have 250 water wheels transporting water from the Yellow River to higher ground. Only 12 of them still work today, but they attract a steady stream of visitors to the area. There, visitors can also take a ride in a sheepskin boat (Figs. 7.1 and 7.2).

The water wheel came to Norway in the 1300s and became a formidable lift in energy supply to grain mills, to saw mills, and for pumping unwanted water from mines. Later, water wheels were also utilized in mills for pulp for paper production, as hammer mills, in iron production, and for similar uses. There are many versions of the water wheel, but in Norway, it was the overshot wheel where water was led to the top of the wheel through a channel that got most of the importance. The current wheel or the undershot wheel, where the water wheel was placed with its lower part in a river or brook or a water channel is another version. Both these alternatives have a wheel with a horizontal axel. So does the breast fall wheel that leads water to the middle of the wheel. We also find water wheels with a vertical axel.

The water wheel is an old invention. According to an energy dictionary, [4] *The Doomsday Book,* written down on order from William The Conqueror in 1087, tells that there were 5624 water wheels in England at that time. Most of these were for grain mills with a current wheel (Fig. 7.3).

The first registered water-driven sawmill was in Germany in 1337 [5]. In 1668, we find about 1200 in Norway that was engaged in export of timber. The intensive timber industry led to razing of the forests, but with new public laws from 1587 to 1860 the activity was regulated to avoid damage of the national resources [5] (Fig. 7.4).

The historians have some disagreement regarding the origin of the water wheel, and it seems reasonable that different varieties of the wheel might have originated various places in the world independent of each other.

Documentation of use of the water wheel in the western world is found from the Greek engineer Philo from Bysanz (Constantinople) about 280–220 BC. Water wheels were also used for pumping of water in Alexandria in Egypt 221–205 BC [6]. The Romans further developed the water wheel,

Fig. 7.1 Old waterwheel in Lanzhou, Gansu Province, Western China

Fig. 7.2 Waterwheels in the ancient city of Lijiang, Yunnan Province, Southwestern China now function mainly as tourism purpose, May, 2009

Fig. 7.3 Old overrun wheel, Margaret River, Western Australia

and several alternatives are described by the Roman architect and engineer Vitruvius.

One of the largest water wheels in the world was used in Kongsberg Silver Mine in Norway, with a wheel of 20 m in diameter and with double blades so that it could run both ways. Water wheels were important in mining to run water pumps, for blowing bellows in the melting huts, and for running the lifts and the crushing equipment.

Without doubt, the water wheel was the forerunner for the more efficient turbine that was developed by Fourneyron in 1827 and was again the forerunner for the turbines that produce our electricity today.

In a global context, however, the renewable energy sources are far less important than the fossil sources for production of electricity. The amount of coal in the world and the prognoses that a new coal power plant will be opened every week until 2020 with half of them in China and India lead to a situation that renewable energy sources will be in minority also in many years ahead. On the other hand, China has taken the lead in the world in investing the

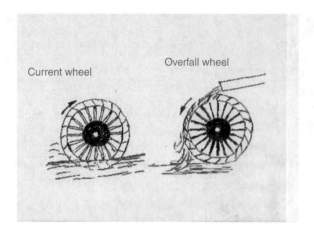

Fig. 7.4 Water wheel (strømhjul = current wheel, Overfallshjul = overfall wheel)

renewable energy and has launched an ambitious target of achieving the ratio of renewable energy to the total primary energy up to 11.4 % by 2015 and 20 % by 2020.

A key challenge for the fossil energy sources is the emission of CO_2 to the atmosphere. In 2002, these sources emitted 21 billion tons of CO_2. Of this the various sources represented [7]:

- Coal 41 %
- Oil 39 %
- Gas 10 %

Nearly 500 coal-fired electricity plants were planned in the timeframe 2009–2019, and more than 700 between 2020 and 2050. With an average living time of 50 years, CO_2 from these plants will continue to heat our planet in hundreds of years to come [7] (Fig. 7.5).

Before we go into some statistics, it might be appropriate to repeat some energy terms:

- *Energia* comes from the Greek language—meaning, activity, power, or strength.
- Efficiency—energy or work per time unit.

   1 W (watt)
   1 kW (kilowatt) = 1000 W
   1 MW (megawatt) = 1000 kW
   1 GW (gigawatt) = 1000 MW
   1 TW (terrawatt) = 1000 GW

- Energy

   1 Wh (watt hour)
   1 kWh (kilowatt hour) = 1000 Wh

Fig. 7.5 Good!—but sun, wind, and hydropower must in a global context accept to play second fiddle to coal as an electricity source in many years ahead. But, the world is in a strong changing mood with respect to energy supply. Teknisk Ukeblad (technical weekly), Oslo, Norway, claims in an article [8] in September 2012 that solar energy the 15. May 2012 represented half of the electricity supply in Germany this single day. The normal for a year is 3 %

1 TWh (terawatt hour) = 1,000,000,000 kWh (approximately the electricity consumption in a year in a city with a population of 50,000) (Fig. 7.6).

Of the renewable sources, hydroelectric power is by far the most important. Geothermic energy has a strong growth, and has great importance in some countries, while tidal water power and wave power still are in a rather modest beginning (Fig. 7.7).

In Norway, we are used to thinking of electrical energy as something that mostly is produced from water power. But this varies considerably from one country to another. As an example, we show the figures from major industrial countries, USA and France from 2009 [9].

	USA (%)	France (%)
Coal	44.9	3.9
Atomic energy	20.3	78.1
Natural gas	23.4	3.8
Hydroelectric power	6.9	11.1
Others	4.6	3.1

Fig. 7.6 Main sources for electricity production in the world [9]

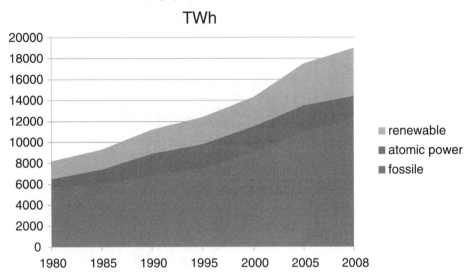

Fig. 7.7 The renewable energy sources [9]

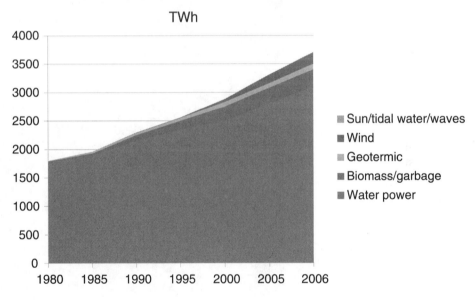

China is still a strong coal-dependent country in the primary energy consumption structure. Hydropower, though takes only 6.8 % of the total, ranks number 1 in the world in the total installed capacity of 220 million kW (Fig. 7.8).

Geothermic Energy

Hot springs have been used for heating, recreation, and bathing since the Stone Age. This is the direct use of geothermal energy which includes balneology (hot spring and spa bathing), agriculture (greenhouse and soil warming), aquaculture (fish, prawn, and alligator farming), industrial uses (drying and warming), residential and district heating.

The oldest known baths we have heard about are from the mountains near China's first capital Xi'an, built by the first Emperor of Qin during the Qin dynasty (221–206 BC) (Fig. 7.9).

We also know that the Romans established bathing facilities, among others in Bath in Somerset in England. The Romans also established hot, mineral-rich baths in a number of places, for example, North Africa, Belgium, Spain, Italy,

Fig. 7.8 Breakdown of China Primary Energy Structure, 2010

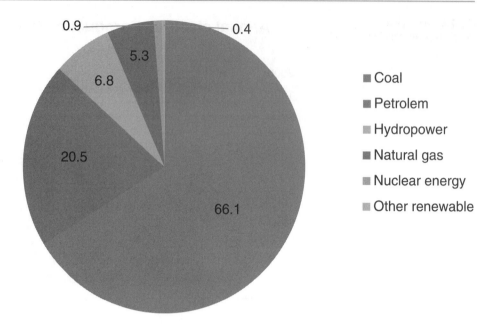

Fig. 7.9 Bubbling hot water. Huaqing Chi palace, near Xi'an, Shaanxi Province, China

France, Czech, and Turkey. All together, there should have been more than 50 such places built by the Romans in various places nearly 2000 years ago.

On Iceland it is told that the famous author, Snorre Sturlason, 800 years ago, had a pond with natural hot water, where he bathed and got inspiration. The story about another poet, Egill Skallagrimsson who lived in Iceland 300 years before Snorre, had to swim 10 km to get ashore on the island of Drangey in North Iceland. When he finally got ashore, he heated himself in a similar pond [10].

In Beppu, Japan where geothermal water and heat have been used in buildings and factories and 4,000 hot springs and bathing facilities, it attracts 12 million tourists a year.

Geothermic energy is the term used, the energy generated from the heat that comes from the heat in the crust of the earth. The energy comes from our own planet instead of from the sun. A below surface mass of rock similar to a medium size mountain (about 520 km^2) holds several hundreds of degrees C which might be enough to cover the need for energy in the world a whole year [11]. The heat is

Fig. 7.10 Countries with geothermic electric power production with capacity of more than 80 MW [13]

COUNTRY	Capacity 2010 (MW)	Percentage part Of national capacity
USA	3086	0,3 %
Philippines	1904	27 %
Indonesia	1197	3,7 %
Mexico	953	3 %
Italy	843	1,5 %
NewZealand	628	10 %
Iceland	575	30 %
Japan	536	0,1 %
Iran	250	
El Salvador	204	25 %
Kenya	167	11,2 %
Costa Rica	166	14 %
Nicaragua	88	10 %

generated in the inner part of the globe when radioactive substances are broken down. You might call it the earth's own natural atomic power station. The heat is then transferred to the surface of the earth.

It is claimed that the first generator to produce electrical energy from the heat of the earth was tested by Prince Piero Ginori Conti in Italy in July 1904. The generator produced enough electricity for four light bulbs [12]. Later, this became the first geothermic electrical power station, and the first modern plant was built in New Zealand in 1958.

Wikipedia, with reference to International Geothermal Association (IGA), tells about geothermic power stations in 24 countries [13] (Fig. 7.10).

The numbers showing geothermic energy as part of the various countries electricity production might be a bit misleading in comparison to the total energy consumption.

Countries, for example, Iceland and New Zealand, use comparatively large amount of natural hot water for heating, and thereby reduce the electricity consumption considerably. Saved energy from natural hot water for heating in a global scale is expected to amount to twice the energy produced as geothermic electrical power.

Iceland was the first country already in 1930 to utilize warm water from the ground as a distance heating source.

In Iceland, close to 90 % of the housing is heated from natural hot water. The hot water in Iceland is energy-wise utilized with [14]:

48 % for heating
37 % for electricity production
4 % for fish farming
4 % for snow melting
3 % for swimming stadiums
2 % for conservatories
2 % for industrial purposes

In total, this means that 84 % of all energy consumption in Iceland comes from natural hot water (Fig. 7.11).

The hot springs in Iceland are divided into 14 high temperature areas and 800 low temperature areas [12]. With their large amount of renewable energy resources, it is hardly strange that Iceland has the highest consumption of geothermic energy consumption per capita in the whole world.

Another "small" but far greener country on the opposite side of the globe with somewhat similar conditions is New Zealand. According to local sources, geothermic electricity accounts for 13 % of the country's electricity production [15]. In total, there are 25 power plants in the enormous Taupo crater (the world's largest volcano crater) on the north island, with installation capacity of 750 MW. Further 10 similar plants with a capacity of more than 1000 MW are to be planned before 2025 (Fig. 7.12).

Geothermic energy has been utilized in many countries that do not have such relatively large sources as Iceland and

Fig. 7.11 From a geothermal demonstration park in Hverdagerdi in Iceland. Hverdagerdi, located about 50 km from the capital Reykjavik, is totally heated from the geothermal possibilities. Herdagerdi is also well known for their water heated conservatories. The local swimming stadium was the first 50 m competition swimming stadium in the country—outdoor of course

New Zealand. A normal procedure is to drill holes into the ground, deep enough to utilize the energy potential in the difference in temperature.

The heating of the house might be a combination of warm water from the ground, heating from solar energy, and from a heating central. Sometimes the system might be combined with hot water from excess from industry or combustion of garbage. To secure an even running of the system, heating from several sources is normal.

Heating pumps have also been used to utilize sea water from a deeper level with stable higher temperature. Other heat sources, as for example sewer plants, have also been utilized.

China's geothermal energy resources are rich and widely distributed, accounting for 8 % of the total of the world. Based on China's Geothermal Energy Utilization launched recently by Ministry of Science and Technology of China [16], use of geothermal energy for power generation started from early 1970s. The first pilot geothermal power station with a unit power of 0.1 MW was built up in Fengshun, Guangdong Province, Southern China. Then pilot unit geothermal energy power generation of 1 MW in Yangbajing, Tibet, China was successfully operated in September, 1977. Now the total installed capacity there reaches 24 MW and has made an important contribution to Lhasa's power demand.

By the end of 2009, the total installed capacity for direct geothermal energy utilization in China had reached 8898 MW with an annual direct utilization of heat energy close to 7.53×1013 kJ, ranked first in the world. Ground source heat pump application takes the lead of about 5210 MW, followed by bathing and swimming application of about 1826 MW and geothermal heating of 1291 MW.

Typical projects include Yangbajing in Tibet for its geothermal power generation, Beijing, Tianjin, and Xi'an for their geothermal heating, Chongqing for its surface water

Fig. 7.12 Bubbling mud and small geysers in Whaka Rewarewa national park and Maori culture centre, Rotorua, New Zealand

Fig. 7.13 Tengchong Hot Lake, Yunnan Province, Southwestern China. Bubbling water with a temperature of 95 °C have been used to boil eggs by local people. For only 5 min the boiled eggs will be done and ready for sale to the tourists

Fig. 7.14 Yangbajing geothermal field for power generation, located in Dangxiong County, 91.8 km Northwestern Lhasa, Tibet, Western China

heat pump based heating and refrigeration, and Dalian for its sea water heat pump based heating and refrigeration, and so forth (Figs. 7.13 and 7.14).

Waterfall - Energy

In the middle of the 1800s, Frenchman Arstide Berges introduced the expression *The white coal* on electricity from waterfalls [4].

Norwegian electricity is mainly produced from waterfall electricity plants (98 %). In 2006, the total production was close 122 TWh, placing the small country of Norway as number 6 in world as waterfall electricity producer. The waterfall electricity production was in the excess of 120 TWh. About 99 % of the electricity production in Norway comes from renewable sources, compared to, for example, 17 % in EU. In the neighboring countries, Sweden and Denmark, the numbers are around 50 and 30 %, respectively [17] (Fig. 7.15).

Fig. 7.15 Røldal Power Station, designed by Arkitect Geir Grung in 1964. Geir Grung have also designed the station at Nesflaten, in the same power complex. That building is ranged as number 7 among Norway's most important buildings The waterfall power development in Røldal and Suldal in the south west mountains of Norway was among the most challenging in the modern Norwegian power development. The system includes 7 power stations. The main development took place in the years 1963–1968

The total economical potential to be utilized of waterfall electricity in Norway, in 1999, was estimated to 178 TWh/year [18]. The protection plan for the water ways has established that not all of this shall be developed, but it is possible to increase the present production within the order of TWh/year, where half comes from new plants, and half from modernization of the existing plants.

A shortlist of the Norwegian waterfall electricity history could be [11]:

1877: Power for two light bulbs at Lisleby Brug near Fredrikstad

1885: First power station—Laugstol Brug i Skien.

1891: First community power station in Hammerfest. First city with electric street lights

1894: The trams in the city of Oslo was electrified

1924: The first world conference on electricity arranged in London, England

1949: 78 % of the population has electricity at home.

The waterfall electricity production in Norway comes from nearly 750 power stations. Today, these power stations are merged into several larger power companies, where the largest is (in GWh/year) [11] (Fig. 7.16):

• Statskraft	33163
• Norsk Hydro ASA	8500
• E-CO vannkraft	7500
• Sira-Kvina kraftselskap	5960
• Bergenshalvøens Kom. Kraftselskap (BKK)	5811
(continued)	

• Lyse Produksjon AS	5232
• Trondheim Energiverk kraft AS	3033
• Hafslund AS	2800
• Skagerak Energi AS	2733

There are other companies that distribute the electricity. The largest in number of customers and distributed power in GWh/year is [11]:

	Customers	Power
• Viken Energinett AS	381400	10815
• BKK Nett AS	151980	4310
• Lyse Nett AS	100000	3648
• Østfold Energi Nett AS	88409	2094
• Trondheim Energiverk Nett AS	83939	2495
• Energi 1 Nordic AS	80000	
• Vestfold Kraft Nett AS	79702	2133
• Nord-Trøndelag Elektrisitetsverk	74781	2372
• Akershus Energi AS	69000	1968

A challenge for the Norwegian electricity network is that most of it was built in the period from 1950 to 1980, and there is a need for considerable upgrading.

Internationally, Norway's probably most famous power station is Vemork Power Station in the Vestfjord valley, west of the city of Rjukan. The station was built by Norwegian Hydro, designed by the Architect Olaf Nordhagen and finished in 1911. When the station was completed it delivered

Fig. 7.16 Kykkelsrud power station in Glomma (Norway's longest river) in Askim community in Østfold County utilizes a fall of 26.5 m and is owned by the company Hafslund. The power station is rehabilitated and was reopened in April 2011 with a capacity of 40 MW [18]. The power station was originally started 1900 and taken over by the company Hafslund in 1915. Before the reconstruction the station had an effect of 22.5 MW, and an average annual production of 35 GWh. The 17 power stations in the Glomma river gives half of the energy consumption in Norway. In Askim community, we also 3 other important power stations; Solbergfoss I—108 MW, Solbergfoss II—100 MW and Vamma—215 MW

147 MW and for a short time it was the largest in the world. The power from the station supplied the energy to Norwegian Hydro's production of nitrate fertilizers, and later also heavy water. That was the reason to the sabotage action in 1943 against the factory that was situated on the plateau in front of the power station (Fig. 7.17).

Heavy water is also called Dideuterium oxide and has denomination D_2O or ^2H_2O. Also the so-called half heavy water exists, where only one of the hydrogen atoms is changed to deuterium. Heavy water looks like ordinary water, but has a somewhat higher freezing point and boiling point. The density is about 11 % higher than for ordinary water. Heavy water is a tool in production of atomic weapons. That was the reason for the sabotage action against Vemork and the heavy water transport during the Second World War when Norway was occupied by Germany.

- In November 1942, 2 Halifax bombers and two gliding planes failed their mission due to bad weather. They failed to meet the forward group of four men on Lake

Fig. 7.17 Vemork Power Station is today Norwegian Industrial Museum. The "new" Vemork Power Station is located inside the mountain

Møsvann, and in total 42 British soldiers lost their life in a plane crash.
- In Operation "Gunnerside" in February 1943, six Norwegian elite soldiers finally managed to meet the forward party and they sabotaged the factory, resulting in blowing up heavy water cells and destroyed 900 kg of heavy water.
- In the massive bombing in November 1943 with nearly 100 bombers, two bombers managed to hit the hydrogen factory. 22 civilians were killed in the local town of Rjukan.
- In the sabotage of the railway ferry "Hydro" on Lake Tinnsjøen in 1944, the rest of the heavy water was destroyed on its way to Germany, but 18 passengers were killed. (See also Chap. 6. Transport lines.)

The height difference in the fall of water is what is utilized in the electricity production. The water magazine is secured by:

- A dam
- Water intake and a water way.
- A power station and a generator.

In the regulation magazine, the water is stored from periods with strong precipitation, where in Norway mostly is in the autumn and in the snow melting in the spring. The electricity consumption is highest in the winter due to natural reasons.

Dams are a key element in the development of water power, and they might be from a few meters high to several hundred meters at the highest.

The largest dams are the largest manmade structures in the world, but there are many versions of them: massive dams or gravitation dams, arch dams, plate dams, etc.

In earlier times, the dams were constructed to provide water from everything from ice production to milling, sawmills, ironworks and similar, but in the last 100 years dams have mainly been constructed in a massive movement for increased electricity production.

Even if other materials, such as earth and stone are important for dam construction, it is concrete that is the key material for building of dams today. Concrete is by far the second biggest commodity in the world only after water, and has a volume more than twice of the volume of all the other building materials together. Annually, we use a concrete volume of a base area of 1×1 km and with the height of Mt. Everest. What is more interesting is that the two world largest commodities, concrete and water, have intimate relationship due the necessity of water for the hydration of hydraulic binder cement. Nevertheless, we have to harmonize the proper use of water in concrete because it could generally worsen the performance of concrete especially the strength if water was used too much in the concrete. Mixing water consumed for concrete every year in the global context can be more than 2 billion m^3 by simple estimation.

The first dam constructed with concrete as part of the structure was made by the Romans about 2000 years ago. A dam which was also the first item where concrete was utilized in Norway, the Farris dam, located outside the town of Larvik and about 120 km south west of Oslo. A test on the 250 years old concrete was made at the Norwegian Building Research Institute in Oslo in the summer of 1990. The test showed very good results not at least due to a very thorough work procedure at the time of the construction (Fig. 7.18).

Fig. 7.18 The first time concrete was used in Norway was in the Farris Dam in 1764. The main purpose of the dam was to secure the town from more flood accidents

Construction of dams has, in many cases and in many countries, been a debated environment issue. Important factors as damming up of fertile agricultural areas, changes in the local climate, damming up of cultural heritage, moving of people, etc., are topics that engage many people. We are not going to contribute to this discussion, and it would also be impossible to draw general conclusion taking into account the many ten thousand of dams that have been built. A sustainable development will nearly always be a balance and a tradeoff between many arguments and factors. However, it is positive that today more and more we often present consequence analysis made prior to new constructions. Such analysis in a transparent atmosphere is important for our future development.

The discussions around regulations of the water ways led to that the Norwegian Parliament raised the question of a national plan for the waterways in the country. In 1973, the Parliament accepted the first protection plan including 95 water ways. Later this led to more and more advanced planning: Protection Plan II in 1980, Protection Plan III in 1986, and Protection Plan IV in 1993 [5].

According to Chinese sources [19], China has the largest hydroelectric resources in the world, even if the country is number 6 with respect to total water resources—2700 billion m^3, after Brazil, Russia, Canada, USA, and Indonesia.

It is hard to say what the most impressive is with the large and fascinating dam projects that have been accomplished around the world: the majestic structures, the impressive organization behind the solutions, the enormous energy sources they represent, the environmental challenges, the often fantastic stories from the constructions, the challenges in material technology, in particular around the temperature difficulties that have to be overcome, or the enormous concrete volumes involved.

Only in Norway, over 2 600 dams have been registered. Many of them represent some of best in Norwegian engineering art, which are the contributions of Norwegian engineering companies both on the design side and on the contractor side to meeting the various challenges in many aspects.

Over time, some gigantic dam structures have been built around the world. The first of these "mega"-dams was **Hoover Dam** over the Colorado River in USA (1931–1936), on the border between the states Nevada and Arizona. The dam has a height of 221 m, length 379 m, and a thickness of 200 m at the bottom and 14 m at the top. Concrete history was written when it was built, not at least with respect to material technology to handle the temperature challenges. In total—3.33 million m^3 of concrete (or about the annual volume consumed in a country like Norway) was consumed in the structure.

The history of the construction is fascinating reading. Here is a small taste:

The Hoover Dam, also known as the Boulder Dam, is an Arch-gravitation dam in Black Canyon, damming up the Colorado River between the states of Nevada and Arizona, USA. When it was completed in 1936, it was both the largest concrete structure in the world and the key element in the world's largest hydroelectric power station. In both areas, however, it was passed by the Grand Coulee Dam in 1942 [18].

The dam is located 48 km south east of Las Vegas and is still a popular sightseeing object.

The dam forms a 640 km^2 water reservoir called Lake Mead.

The construction of the dam started in 1931 and was completed in 1936, 2 years ahead of the time plan.

After many previous discussions ahead, a government commission was formed in 1922 to decide what to do with the Colorado River that had many flood catastrophes with serious damages. The river conserves many states: Arizona, California, Colorado, Nevada, New Mexico, Utah, and Wyoming. The Governors from these states met with the representative of the government representative Hoover in January 1922. Hoover was then Secretary of Commerce under President Warren Harding. The commission agreed on November 24, 1922 on an agreement which among others meant a splitting of the planning in an upper and lower part. This also became the bases for the planning of the Boulder Dam. The agreement was named "The Hoover Compromise."

Already in 1922, the first proposal for the dam was put forward in the US Congress. However, it was not until December 1928 that both the Congress and the House of Representatives approved the bill. The President signed the law on December 21, 1928. At that time, it was President Calvin Coolidge who was the President. Preparations for the construction started in July 1930.

The construction work had a lot of challenges. An important argument for speeding up the work was to secure working places after the great economic depression. A new city was built for the workers, Boulder City, but this city was not ready when the first workers arrived early in 1931. Dissatisfaction with the intermediate living quarters was the reason for a strike among the workers on August 8, 1931. The work with Boulder City was speeded up, and moving from the intermediate "city"—"Ragtown" which was ready during springtime of 1932. Gambling, alcohol, and prostitution were forbidden in Boulder City. Even today, Boulder City is one of two cities in the state of Nevada that does not allow gambling. Sale of alcohol was allowed in 1969.

Officially, 96 workers died during the construction of the dam, many due to the contaminated air in the tunnels. The unofficial numbers say that there were 112 deaths. A rather special and a bit popular story claims that the first death was the surveyor J.G. Tierney, who drowned on his job in the start of the construction. His son P.W. Tierney is said to be the last that died during the construction of the dam 13 years later [20].

The concrete work on the dam started on June 6, 1933 and gave a series of challenges, especially on the temperature issue. A number of technologies were developed to solve the problems, and many of them have become standard procedures during dam construction later. Among others, there were installed 1 in. (25 mm) cooling pipes in the concrete. These were cut and injected with mortar later on. All concreting took place in trapezoid casting sections. The 3.33 million m^3 of concrete in the dam should have been enough for a two-lane motorway across USA from San Francisco to New York.

Herbert Hoover played an important role in the building of the dam. That is why the dam got his name. But there has been considerable controversy regarding the name. Originally the dam was called Boulder Dam. Then Hoover became President of USA on September 17, 1930, his Secretary of the Interior, Ray Lyman Wilbur, proclaimed that the dam should be renamed to Hoover Dam. It was not unusual that large dams got a president's name. This was officially approved by the Congress on February 14, 1931. In 1932, Hoover lost the president election to Franklin Delano Roosevelt. Roosevelt's Secretary of the Interior, Harold Ickes then on May 8, 1933, issued the memorandum that the original name, Boulder Dam should be used on the dam. Roosevelt resigned as President in 1945 and Ickes retired in 1946. On March 4, 1947, a proposal was made from a congressman from California that it again should be Hoover Dam. This was then approved by the two chambers on April 23 and 30, 1947.

At last, a small story was mentioned in the September issue of Fortune Magazine in 1933 [21];

Arthur Powell Davis was an engineer who started to get interested in the dam project as early as in 1902. His uncle, John Wesley Powell, did the first investigations of Grand Canyon already in the 1860–1870s. Arthur Powel got his education as civil engineer from Columbian University (now George Washington University) in 1902. For 14 years, he had been fighting for construction of the dam through his work at the potential building responsible—US Bureau of Reclamation. Controversies between Davis and his leaders, however, became so big that he in 1923 delivered his resignation as director and chief designer in the bureau. Disillusioned he returned to California, where he worked with local hydraulic projects. Later he worked in Turkistan as Soviet's chief consultant for irrigation projects.

The work on Boulder Dam continued for 10 years without Davis. Davis lived in California and his health was reduced. Then in July 1933, Davis got his satisfaction by getting hired again as chief consultant for the dam. Davis was then 72 years old, and his health did not allow so intensive work anymore.

The next large giant dam was **Grand Coulee Dam** (1933–1942), located in the state of Washington, damming up Colombia River. The dam is 68 m high, and has a length of nearly one English mile (1592 m). In 2008, the dam was still a very important component in the fifth largest electricity production in the world. In total, 9.13 million m^3 of concrete were used in the dam.

In the years 1970–1984, a new giant step was taken on the "dam-record" ladder by the construction of **Itaipu Dam** on the border between Brazil and Paraguay. The dam is part of what at the time became the largest power station in the world, and which in 2008 produced 94.7 TWh [22]. In the power station, 20 generators were installed, where 10 produce electricity for Brazil and 10 for Paraguay. The giant station makes Paraguay the only country in the world being 100 % supplied by hydroelectric power [23] (Fig. 7.19).

The construction of the Itaipu dam over the Parana River started in January 1970, and the opening took place on May 5, 1984 after a spending of 19.6 billion USD. The dam solution is rather four dams linked together, one earth filling dam, one rock filling dam, and two concrete dams. The total length of the dam is 7235 m with a height of 225 m at the highest [23].

The construction of the dam involved several agreements:

- As a foundation for the construction is the agreement between Brazil and Paraguay from July 22, 1966 about investigations for the project.
- In 1973, the two countries signed the agreement for the building of the power station with a running time to 2023.
- On October 19, 1979, Argentina, Brazil, and Paraguay signed an agreement about the regulation about the water

Fig. 7.19 The Itaipu Dam on the border between Brazil and Paraguay

level in the river to secure against floods in River Plate that runs through the capital of Argentina, Buenos Aires.
- In 2009, Brazil accepted an adjustment of the payment for electricity to Paraguay.

The filling of the reservoir behind the dam started on October 13, 1982. Due to a period of heavy rain and flooding, the water was rising 100 m and reached the flood outlets already on October 27. The reservoir behind the dam is 29 billion m^3 and covers an area of 1.35 million km^2.

American Society of Civil Engineers in 1994 selected Itaipu Dam as one of the modern seven wonder structures of the world.

Megaproject like this has a large effect on the electricity production. Half of the hydropower capacity of Itaipu representing Brazil's part is comparable to burning of 69 000 m^3 of petroleum every day. Nevertheless, some of the backside of the medal is the vulnerability of such projects. During a gale on November 10, 2009, the masts in three high current electricity lines between the dam and Sao Paulo were destroyed. This led to a massive power cut in Paraguay and Brazil. Paraguay was completely darkened in 15 min, and the megacities Sao Paulo and Rio de Janeiro were without electricity for over 2 h. Fifty million people were affected by the power cut.

The Three Gorges Dam, across the Yangtze River near Sandouping in Yichang, Hubei province in China, is the world's largest concrete structure and hydropower project till now.

The dam is not only the foundation for a gigantic power plant, but is of uttermost importance as flood securing magazine for the Yangtze River. The river, through history, had a number of catastrophic floods [24];

- 1931: 130 000 km^2 were flooded with effect on 28.55 million people, 145 000 died.
- 1935: 89 000 km^2 flooded, effect on 10 million people, 142 000 died.
- 1949: 1.81 million ha of agricultural land area flooded with effect on 8.1 million people, 5699 died.
- 1954: 3.18 million ha agricultural land area flooded, with effect on 18 884 million people, 33 169 died. The Beijing–Guangzhou railway line closed for 100 days.
- 1998: All of China participated in the fight against the flood for months. 6.7 million people and several hundred thousand soldiers participated actively. 1526 died.

The Yangtze River, the largest one in China with the main stream more than 6300 km long, is also known as Chang Jiang (long river) and is the third longest river in the world after the Nile and Amazonas. (See more in Chap. 5. Rivers and lakes.) From the origin of the river—the Qinghai-Tibetan Plateau to the entrance to the sea in the east in Shanghai, the big drop of 6000 m and abundant water flow make the Yangtze River the richest hydropower resource in China. So far about 100 hydroprojects have been built along the river basin. Gezhouba hydropower station (installed capacity 2.7 million kW) located 38 km downstream of the Yangtze River from Three Gorges dam is the earliest one along the main stream, which was started in 1971 and completed in 1988.

The large rivers are of great importance for the internal transport systems in China. An important advantage with the Three Gorges dam is the special locks. Together with the dam, a gigantic lock system was built with chambers of 132 × 23.4 m. This allows 3000 tons barges and boats to be lifted up all together 113 m, and makes it possible with river transport from the sea up to the upstream megacity Chongqing in the Sichuan province.

The first to come with the idea to dam up the Yangtze river is said to be the father of modern China, Sun Yat-sen in 1919 [25]. After another large flood, Mao Tse-tung ordered an investigation for a dam to be started in the 1950s.

There have been critical voices against dams in both China and in other countries due to the environmental consequences. With close to 1/3 of the representatives of the National People's Congress voted for objection or waiver, the resolution was made on April 3, 1992 by Chinese government for construction of the Three Gorges mega hydropower project. Chinese authorities claim that in a holistic sustainability perspective, there is hardly any doubt that building of the dam has been a wise decision.

The work on the dam started on December 14, 1994, and the dam itself was finished in May 2006. The dam is 181 m high and has a length of 2 309 m, a width in the bottom of 115 and 40 m at the top [26]. The mega hydropower project was fully completed in 2009.

With an installed capacity of 22 500 MW, when all the 32 main turbines are running, the power production is comparable to 18 atomic power stations, or 1/9 of the power consumption in China [24] (Fig. 7.20).

The second longest river in China—Yellow River, historically the Mother River of China is also a catastrophic river. During the more than 2500 years history from early Qin Dynasty to 1949, the river burst for over 1590 times and diverted in a large scale for 26 times [27]. This is the fact as local folklore said in ancient China on Yellow River, "Divert once every 100 years, Burst twice every 3 years, and Disaster every year." The evolution of Chinese civilization has been carried on with continuous combat with the river. It

Fig. 7.20 The Three Gorges Dam was built from 1993 to 2009. The power station equipped with a total installed capacity of 32 × 700 MW produces every year 100 billion kWh. The huge concrete gravity dam construction consumed 28 million m^3 of concrete, where 16 million m^3 for the dam itself [26]

is only after the funding of the People's Republic of China since October, 1949 which creates a new epic of eradicating the disaster and utilizing the river, as are witnessed below:

- July 18, 1958, Chinese government issued the programming on eradicating the flooding disaster of the Yellow River and developing water conservation projects;
- August, 1958, the first large hydropower station—Qingtongxia was started to be built with its first unit commissioning in 1967;
- September, 1958, China's first hydropower station with an installed capacity over 1 million kW—Liujiaxia was started, and it was commissioned in December, 1974 with an installed capacity of 1.225 million kW;
- April, 1998, Yuantou hydropower station with a height of 3980 m above the sea level and the first one from the origin of the Yellow River was started to be built and completed in December, 2001 with an installed capacity of 2500 kW;
- Early 2004, the Yellow River was dammed as a start for constructing the largest (700 MW × 6 units) and highest (maximum height of the dam: 250 m) hydropower station of the river basin. The first 5 units were successfully commissioned in August 2010 and the whole project was completed in November 2011.
- Totally there are 43 water conservation projects along the main stream of the Yellow River, 19 hydropower stations among which have been built. Seven of the rest are under construction and 21 stations are in planning. The reason why only five projects are located in the middle and downstream river is due to the high silt content in that segment [28].
- Before 1949, the irrigation along the Yellow River was only 12 million mu (1 μm = 666.7 m^2), while this becomes 100 million mu with a total irrigation benefit of 400 billion RMB Yuan (Fig. 7.21).

As mentioned above, a number of dam projects have been controversial in more than one way.

The Rybinski reservoir in Russia starts about 260 km north of Moscow and stretches further 140 km against the north. This water magazine has an area of 4 500 km^2, and from the time when it was built up to many years later was the largest in the world, being bigger than Rikon in Uruguay and Kentucky in USA. The Rubinsky reservoir is located where the rivers Volga, Sheksna, Mologa, and hundreds of other smaller rivers meet. The reservoir was built in the 1930s and 1940s and was completed in 1941 to secure stable navigation conditions in the shallow water way between Moscow and the Baltic Sea (St. Petersburg). Before this 40 rapids made the sailors to pull the bout forward with ropes. (Even if there is no need for the barge pullers any more, there are still a number of artist that have songs about them on their repertoire.) As many of Stalin's projects, it was built without much thought about the nature and the people affected by it. Still, 70 years later, there is lack of oxygen in the water due to all the forests and the villages that were put under water, and there is hardly any fish in the reservoir. Most of the reservoir, 2590 km^2 in total, has a water depth of less than 6 m.

At the southern end of the Rybinski reservoir, we find the city of Uglich. In Uglich there is a power station damming up Volga further up in what is called the Uglich reservoir. This is a reservoir that is 143 km long but only 500–1000 m wide. The reservoir covers 249 km^2. In the reservoir, among others a number of churches have been covered with water. In the town of Kaliazin, the tower of the St. Nicholas church is still sticking up over the water. In Uglich the dam is located just before coming into the Rubinski reservoir.

Fig. 7.21 Reservoir of Liujiaxia Hydropower Station covering an area of over 130 km², located in the main stream of Yellow River in Yongjing County, Gansu Province, China, where distinct boundary of the Yellow Silt water mixed with the clear water after deposited in the reservoir. *Photo* Hao Sui

Fig. 7.22 The dam and the power station in Uglich. The administration building to the far left, and a renovated church in between

Uglich is a town with 40 000 people and one of the oldest in the Jaroslav region. A monastery, churches and 1500 and 1700 hundreds buildings were demolished to make the foundations for the big power station (Fig. 7.22).

Another example of giant dam project that has had large environmental consequences is the Aswan dam in Egypt.

The first trial to build a dam near Aswan started as early as in the twelfth century. Then in 1898 the British started building the low Aswan dam, about 7 km further down from where the high dam is located today. This dam was opened on December 10, 1902. In 1946 this dam was nearly destroyed by flood. This initiated the plans for building the high dam further up from the old one [29]. More speed in plans came with the Egyptian revolution in 1952.

The high dam at Aswan was constructed in the years 1960–1970

The purposes or the needs for building the dam were several:

- Reduce the flood damages in Nile valley downstream from the dam.
- Increase the agricultural and industrial potential of Egypt.

- Improve the shipping traffic in Egypt.
- Increase the electrical capacity. Originally the power station of the dam was intended to double the electrical capacity of Egypt. The power station has a capacity of 2.1 GW, and the need in Egypt in 1976 was 3.8 GW. Later the need has increased fivefold. In 2020, the expected consumption in Egypt is estimated to 26 GW [30].

Damming up the Nile created the 270 km long Nasser Lake. The irrigation systems from here and to the agricultural land downstream have had a considerable positive effect on the agricultural production in Egypt. The Nasser Lake has also created an important local fishing industry.

The environmental problems with the dam have created considerable discussions and headlines [31]:

- About 100 000 people got their homes and farmland flooded.
- Important old memories from the past were threatened by the damming. Large projects were initiated to save the most important ones. An example is Abu Simble Temple that was moved to a higher level, a gigantic saving operation, and an impressive engineering work.
- The Nile brings along considerable quantities of silt. Large quantities of silt are deposited on the bottom of the Nasser lake and give challenges
- The silt was previously deposited along the Nile, and was an important soil improver for the agriculture. This now has to be compensated by artificial fertilizers if the soil shall have similar fertility.
- The evaporation of the water in the Nile seems to have increased. The delta area at the mouth is sinking and gets swampier. This seems to have negative effect on the fish quantity in the delta.

Small Power Stations

In Norway, small power stations are normally categorized in three categories.

- Micropower stations: with effect of less than 100 kW.
- Mini power stations: with effect between 100 and 1000 kW.
- Small power stations: with effect 1and 10 MW.

The Government agency, NVE (Norwegian Waterway and Electricity Directorate), can give concessions to power stations less than 10 MW, while the Norwegian Parliament decides concessions for larger power stations. Evaluating the potential for small power stations to be established, it is estimated to about 24 TWh, while the potential for increased power from upgrading of old stations is about 15 TWh [32]. Of the potential of 24 TWh, NVE claims that it is realistic to realize 5–10 TWh in the next 10 years (from 2004).

An interesting point is that the many small power stations can be established with relatively small effects on the nature.

Small hydropower resources in China are very rich and widely spread in more than 1600 counties. The statistics [33] show that by the end of 2006 China had built 46 489 small hydropower stations with an installed capacity of 44.9 GW, accounting for 34.9 % of the country's total hydropower. The dynamic development of small hydropower stations remarkably reduces the power cost and highly meets the increasing demand for electricity in the vast rural areas.

Other Water Power Solutions

There are a number of other hydropower alternatives. Some of them have resulted in considerable power stations, but in a global scale they are of relatively small importance in the total picture.

Examples are: river stream power stations, wave power stations, ocean current power stations, ocean temperature power stations, and tidal wave power stations.

The formidable forces established when large masses of water are in movement have inspired developers all over the world, and quite a few of their ideas have been realized. The Norwegian newspaper, Aftenposten on August 17, 2009 mentions among others [34]:

- 40 000 British houses will get electrical power from tidal wave turbines in 2011, and 40 tidal wave turbines might be placed in an area north of Scotland.
- The ocean wave company Pelamis Wave Power opened a station outside the coast of northern Portugal in September 2008.
- The company Fobox has developed their own ocean wave technology and is planning to use their technology in a wave farm outside Cornwall in England.

This type of energy from the sea might be utilized in many alternatives. An example is registered from the technical journal, Teknisk Ukeblad, Norway in September 2012, that reported in planning a bridge across the Bokna fiord, forces against the bridge structure from wind and water are planned to produce electrical energy.

Common for all the ocean power systems is that they give large challenges with respect to: cost, tolerances versus weather problems and corrosion risk.

Tidal Wave Electrical Power

The largest tidal wave power plant was constructed outside Rance in Bretagne in France in 1966. The turbine effected is 240 MW. In spite of the high development costs, the costs have been paid back, and the energy production costs are lower than for atomic power [35].

This technology is looking for narrow sounds that might give higher efficiency. The building costs are high, and the technology is only interesting when there is considerable difference between low tide and high tide.

There are several systems that transform changes in ocean level between low and high tide to electrical energy:

- Utilizing the height difference

Water collected in a magazine during high tide is led through a turbine at low tide.

- Ocean mills

Where the water is pressed to and from between high tide and low tide is located a windmill looking equipment with rotating blades, which produces electrical energy.

The interest for tidal water power is global. This is well illustrated by Wikipedia [36] mentioning among others tidal water power plants in (both commercial and prototypes):

- Busum tidal wave power station, Germany, as the world earliest pilot tidal power station in 1913
- East coast of China, more than 100 small tidal wave power stations were built in 1950s yet most of them stopped operation soon after completed due to varying reasons, e.g., improper site selection, poor design or equipment, etc. For instance,
  - Rongcheng in Weihai, east coastal town of Shandong Province, China, destroyed soon after it was completed due to the poor quality, 10 kW, 1958 [37].
  - Shashan tidal wave power station, Zhejiang Province, China, 40 kW, 1965
  - La Rance, Saint-Malo, France, 240 MW, 10 turbines, 1966
  - Murmansk, North of former USSR, 800 kW, 1968
  - Jingang tidal wave power station and Baishakou tidal wave power station, both started construction in 1970, Rushan in Weihai, east coastal town of Shandong Province, China. The former stopped operation 3 years after completion, and the latter with its first unit commissioning in 1978 and full operation of the 6 units with a total capacity of 960 kW in 1987, withdrawn from operation due to local government plan in 2007.
- Projects later built include:
  - Jindo Udolmok, South Korea, 90 MW, 2009–2013
  - Annapolis Royal Generating Station, Fundy Bay, Canada, 19 MW, 1984
  - Jiangxia, Zhejiang Province, China, 3.9 MW, 6 turbines, 1980, full operation in 1985
  - Xingfuyang, Fujian Province, Southern China, 1280 kW, 1989
  - Kilaya Guba, Murmansk, Russia, 1.7 MW, 2 turbines, 2004
  - SeaGen, Stratford Lough, North Ireland, 13 MW, 2008
  - Sanmen tidal wave power station, Zhejiang Province, China, 20 MW, 2009

Wave Power

The idea about utilizing the energy from the ocean waves is claimed to originate from France more than 200 years ago. The enormous interest in the possibility worldwide can be illustrated by the fact that more than 1000 patents on the topic have been issued [38].

The size of the waves depends on both the wind velocity and the working distance, and their continuity is also an important efficiency factor. The wave motions are surface phenomena. Accordingly, the utilization of the energy will be likewise.

To transform the wave energy to electrical energy might be done with different technology systems:

- Moving water column

The wave's motion is transferred to a vertical pipe or a column with air that has an opening in the bottom. The wave action forces air to be pressed through a turbine. The turbine rotates and produces electrical energy. A somewhat similar system has a water engine located on the sea bottom.

- The channel system

The waves coming against the shore press water through a channel to a magazine at a higher level. When the water runs back again, it is led through a turbine that produces electrical energy.

- Wave dragon

A large complex horizontal arm system, looking like a reflector that leads waves to a ramp that leads the wave to a magazine in the back. The water runs through turbines as it is led back to sea level.

Fig. 7.23 The energy consumption in the world in 2010 [43]

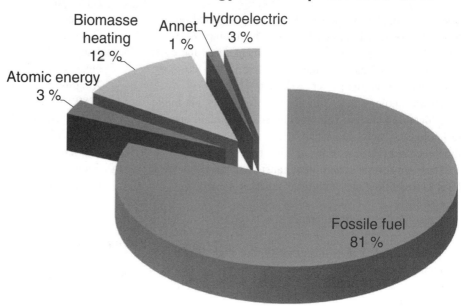

- Water piston

The waves are moving a vertical pipe with a water piston, again running a turbine.

Ocean Heat Power

The temperature difference between sun heated warm surface water and water in depth with lower stable temperature can be utilized, which is called ocean thermal energy conversion (OTEC). However, it is only in tropical waters that the temperature difference is of such a scale that this system could be used commercially.

Test in using these phenomena, which might have considerable possible energy potentials, has been discussed in more than 100 years. OTEC theory, according to Wikipedia [39], was first developed in the 1881 by a French physicist, Jacques Arsene d'Arsonval, who proposed tapping the thermal energy of the ocean. His student, Georges Claude, built the first OTEC plant in Matanzas, Cuba in 1930, which generated 22 kW of electricity with a low-pressure turbine. The plant was later destroyed in a storm.

In Hawaii, the Natural Energy Laboratory was established in 1974, and in 1999 they built a power plant of 250 kW. A much larger scale pilot project of 10 MW via OTEC has also been under construction in Hawaii by American Aerospace Corporation in cooperation with Makai Ocean Engineering. The project is expected to be commissioning in 2015 [40]

Japan is also a major contributor to the development of OTEC technology [41]. A 100 kW closed-cycle OTEC plant was successfully built on the island of Nauru in 1970 and became operational in 1981, producing about 120 kW of electricity.

A joint plan by Lockheed Martin and Reignwood Group has been launched in 2013 to build world largest 100 MW OTEC project in China [42]. The pipe for this project with a diameter of 10 m and length of 1000 m in order to extract enough cool water about 5 °C under the sea is a big challenge, which costs 1 billion USD.

Ocean Current Energy

Some people think that utilizing the enormous energy potential in the ocean current might be an interesting option. Experts have, for example, claimed that the energy in the Golf Stream Current is more than 20000 larger than the energy in the Niagara Falls. The first ocean current power turbine was made and operated on trial scale in Italy in 2001 with a capacity of 40 kW. So far, this alternative is on the testing stage, but interesting research is going on both in Europe, America and Asia.

Water—The Giant Energy Consumer

It is important not to forget that considerable amounts of energy are needed to provide the water we consume. In the UN water report [43], it is estimated that 7%–8% of all energy produced in the world is used for pumping up ground water, for pumping lines and for irrigation purposes. In some developing countries this figure might go up to 40 %. At the same time, the report states the number of people in the world without electricity accounts to 1.4 billion, or 20 % of the population on the earth.

The same report, with reference to IEA (International Energy Organization), expects that 5 % of the road transport in the future (2030) will be driven on bio fuel. To produce this fuel, 20 % of the water consumed in agriculture might be needed if we do change our working methods. (EU has goal of 10 % of the road transport on bio fuel in 2020. China has a goal of 15 % bio fuel nationwide before 2020.) This will put further pressure on an already unsatisfactory access to water, and also mean a considerable increase in the need for electricity. This problem will in particular be difficult and problematic in areas as West Africa.

It is not without consequences to touch the ecological balance in the world.

You might be thoughtful by the fact that hydroelectric power in total represents 3.34 % of the energy consumption in the world [43], while at the same time the world consumes 7%–8% of the energy production to provide us with water. **In other words, the world hydropower production is only half of what is needed to provide us with enough water** (Fig. 7.23).

References

1. Mary Bellis, Water wheel, Part I: http://inventors.about.com/library/inventors/blwaterwheel.htm.
2. Access on May 10, 2014. http://hua.umf.maine.edu/China/astronomy/tianpage/0012ZhangHeng6539w.html.
3. Chen Ledao, Historic Data Study - Process and history of waterwheel in Lanzhou based on archive records. Archives, No. 3, 2008 (in Chinese).
4. Arneson Steinar: *Energi Lex 2001*, Villrose Norsk Forlag, Oslo, Norway, 2001.
5. Internet 12.10.12: Wikipedia: *Water wheel*, http://en.wikipedia.org/wiki/water_wheel.
6. Thoresen Carl A.: *Port designer's handbook*, Thomas Telford Ltd., London 2010.
7. Flannery Tim: *Værmakerne (The Weather Makers)*, H. Aschehoug & Co, Oslo, Norway, 2006.
8. Teknisk Ukeblad, Leder: *Se til Tyskland! (Look to Germany)*, Teknisk Ukeblad 27. Oslo, Norway, September 2012.
9. Internet 18.04.11: http://en.wikipedia.org/wiki/Electricity-generation.
10. Internet 23.09.12: *How Geothermal Energy is used in Iceland*, http://waterfire.fas.is/GeothermalEnergy/GeothermalEnergy.php.
11. SFT: *Veileder for klassifisering av miljøkvalitet i fjorder og kystfarvann (Manual for classification of the environmental quality in fiords and costal areas)*, SFT (The Government Pollution Agency), Oslo, Norway, February 2008.
12. Icelandair: *Iceland—Information Guide 2006–2007*.
13. Internet 22.09.12: http://en.wikipedia.org/wiki/Geothermal_energy.
14. Internet 23.09.12: Orkustofnun: http://www.nea.is/geothermal.
15. Internet 23.09.12: New Zealand Geothermal Association, http://www.nzgeothermal.org.nz/elec_geo.html.
16. China's Geothermal Energy Utilization, Ministry of Science and Technology of China, pp 1–38.
17. Internet 22.09.12: Miljøverndepartementet (The Norwegian Department of the Environment). http://www.Regjeringen.no/nb/dep/md/dok/regpub/stmeld/2006-2007/stmeld-nr-34.
18. SINTEF: *Energikilden—En guide til kilowattens rike (The Energy Source—A guide to the world of kilowatts)*, Tapir Akademisk forlag, Trondheim, Norway, 2000.
19. Zheng Ping: *China's geography*, China Intercontinental Press, ISBN 7-5085-0914-5/K-751.
20. Internet 25.11.09: Hoover Dam, http://en.wikipedia.org/wiki/Hoover-Dam.
21. Internet 24.11.09: Fortune Magazine 1933. http://www.usbr.gov/lc/hooverdam/History/articles/fortune1933/html.
22. Frithjof Gartmann: *Sement i Norge—100 år (Cement in Norway—100 years)*, Norcem AS, Oslo, Norway, 1990.
23. Internet 18.04.11: http://en.wikipedia.org/wiki/Itaipu-Dam.
24. http://edition.cnn.com.SPECIALS/1999/china.50/asian.superpower/three.gorges/.
25. Internet 18.4.11: http://en.Wikipedia.org/wiki/Three-Gorges-Dam.
26. Tongbo Sui. High Performance low heat Portland Cement—High Belite Cement, Preparation & Performance, Application and Perspective. 2007 International Workshop on Cement & Concrete Technology for Sustainable Development, Beijing, Nanjing and Lhasa, August 2007, China Building Material Society, Beijing 100024, China.
27. Access on March 2, 2014. http://baike.so.com/doc/4951143-5172588.html.
28. Access on March 2, 2014. http://blog.sina.com.cn/s/blog_5938096a0100088n.html.
29. Internet 23.10.12: Wikipedia: *Aswan Dam*, http://en.wikipedia.org/wiki/Aswan-Dam.
30. Holm Yngvar: *Den store boken om Amerikabåtene. Nasjonens maritime stolthet*. Edvarde'en Forlag, Bergen 2004.
31. Internet 23.10.12: *Environmental Impact of the Aswan High Dam*, http://www.mbarron.net/Nile/envir_nf.html.
32. Norsk Betongforenings Miljøkomite The Norwegian Concrete Society, Environment Committee: *Økt fokus på Miljø og Miljøutfordringer—Nye mulighet for betong (Increase focus on the environment and its challenges—New possibilities for concrete)*. Norsk Betongforening, Oslo, Norway 2010.
33. Current Status and Future of Small Hydropower Plant, December 5, 2007. http://wenku.baidu.com/view/9ac4494f852458fb770b5617.html?re=view.
34. Andersson Atle: *Vindstille i Norge (Wind quite in Norway)*; (Bergens Tidende) Aftenposten, Oslo, Norway, 17.08.09.
35. Teknisk Ukeblad: *Bruer kan bli kraftverk(Bridges can be power stations)*, Teknisk Ukeblad 29/12, Oslo, Norway.

36. Internet 30.09.12. *Tidevannskraft (Tidal Power)* http://no.wikipedia.org/wiki/Tidevannskraft.
37. Lu Shan, Taizhou Evening News, January 23, 2011.
38. Internet 30.09.12: *Bølgekraft (Wave Power)* http://home.no/vannenergi/bolgekraft.html.
39. Access on March 2, 2014. http://en.wikipedia.org/wiki/Ocean_thermal_energy_conversion.
40. Access on March 2, 2014. http://www.ditan360.com/Finance/Info-98980.html.
41. Bruch, Vicki L. (April 1994). An Assessment of Research and Development Leadership in Ocean Energy Technologies (PDF). SAND93-3946. Sandia National Laboratories: Energy Policy and Planning Department.
42. Access on March 2, 2014. http://intl.ce.cn/specials/zxxx/201304/19/t20130419_24306973.shtml.
43. Internet 14.03.13: *Total World Energy Consumption 2010*, http://upload.wikimedia.org.wikipedia/commons/6/67/Total_World_Energy_Consumption.

Erosion

8

Erosion is a natural process in nature where earth and rock are worn away and moved to other locations mainly by wind, water and their combination, among which water is a very forceful and efficient tool for water erosion. The large volumes of water and the impressive forces in water and ice in motion, give over time important and marked changes in our topography.

An interesting phenomena for rivers on the earth is that the right bank of the rivers in the Northern Hemisphere appears much steeper due to the accumulation of slightly higher erosion for a long time, while in the Southern Hemisphere the left bank of rivers looks steeper [1]. This can be mainly attributed to the Coriolis Effect arising from the earth's rotation—an effect that as the earth spins in a counterclockwise direction on its axis anything flying or flowing over a long distance above its surface is deflected. The direction of deflection from the Coriolis Effect depends on the object's position on Earth—objects deflect to the right in the Northern Hemisphere while in the Southern they deflect to the left. The mathematical expression for the Coriolis force appeared in 1835 in a paper by French scientist Gaspard-Gustave Coriolis in connection with the theory of water wheels. Early in the twentieth century, the term Coriolis force began to be used in connection with meteorology [2].

Some of the most important impacts of the Coriolis Effect in terms of geography are the deflection of winds and currents in the ocean.

Human activity, however, often accelerates this process. Some researchers are of the opinion that they can record a 10–50 folds of increase in the erosion due to human activity.

Precipitation, rivers and brooks, glaciers, avalanches, waves, frost, and ground water transport are all typical causes for erosion.

Materials like soils, rocks, and eroded sediment may be transported just a few millimeters, or for thousands of kilometers. When materials are taken away from its original position by water or glaciers, glacier calving or stone slips, or through evaporation, we call this ablation. For erosion from tidal waters, surfs or waves, we call it abrasion which is caused either directly by water or indirectly through water.

Excessive or accelerated erosion as one of the most significant environmental problems worldwide exerts lots of impacts on both natural environment and human life, such as the problems of decrease in agricultural productivity and collapse of local natural landscapes or ecological system, and eventual desertification, sedimentation of waterways, as well as sediment-related damage to roads, houses, etc.

Intensive agriculture, increasing deforestation, excessive roads in particular impermeable concrete or bitumen roads, anthropogenic climate change and fast urban expansion are among the most significant human activities imposing their effects on stimulating erosion. Examples of human activity that lead to increased erosion are many, e.g., the increased urbanization leads to less porous ground, which again lead to faster collection of the water, higher volumes and increased water speed in the runoff, and thus enhanced erosion. In a natural environment possibly only 10 % of the precipitation goes to direct runoff, while the rest is absorbed in the ground buffer, and is used for longer time to recipient, consumed in the vegetation, or will evaporate.

Typical values for absorption or runoff to recipient is

Surface type	Ground absorption (%)	Runoff to recipient (%)
Unspoiled nature	95	5
Agricultural area	70	30
Suburbs	30	70
City centers	5	95

Increased erosion, in worst consequence, results in flooding. The consequences of flooding also increase because it more and more happens in larger concentration of people.

Outside the urban areas, human make encroachments and use of the area can also increase the possibility for flooding.

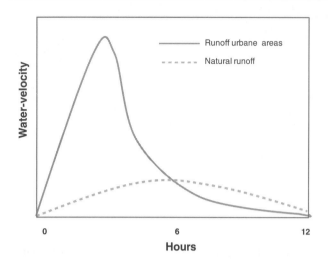

Fig. 8.1 In urban areas the water runoff goes faster than in natural surroundings. With nonabsorbing surfaces, the damping of the runoff effect from vegetation and porous ground is drastically reduced

Examples of this are flattening of the landscape, intensive lumber activity, closing of brooks, etc.

Due to the increased damages, a better handling of surface water has got greater attention in most countries in recent years. This leads to increased awareness in planning, and new solutions and ideas coming into use (Fig. 8.1).

Permeable concrete slabs have a growing popularity because of its ability both in delaying the water runoff to the recipient and in reducing the surface water in typical walking areas (Figs. 8.2, 8.3 and 8.4).

Other typical tools to delay the runoff in urban areas are

- Green roofs (Reports tells about reduction up to 50 % in the runoff from these areas)
- More plants and trees
- Delaying ponds and lakes.
- Delaying basins and pipe systems below ground.
- Increased capacity of the runoff piping systems.
- Movable delaying edges in the street runoff.
- Shallow, low grade grass covered permeable instead of hard gutters.
- Reduction in number of manholes
  Etc.

The combination of climate change with warmer weather in dry periods and more heavy precipitation in the rainy periods combined with increased urbanization with a tendency to getting denser surface in large areas have increased the flooding problems considerably.

As mentioned in Chap. 1, water is the life giving and the origin of all matters. Yet water can be disastrous if it comes too much, too fast or both, forming the destructive flood for instance. From flood myths worldwide in the ancient time to the modern world today, the history of human civilization has been progressing with the combat with flood disasters. Statistics of China's record on history of flood gives 1092 large scale flooding from 206 B.C. to 1949. International record from Emergency Events Database also reports 2565 floods in 172 countries from 1972 to 2005. India and China among which with 141 and 127 floods occurring respectively are the most suffered in the world. Typically in 1998, China, India and Bangladesh are three countries in the world heavily hit by flood which each caused thousands of people died and millions of people affected.

EU and Norway had decided on a flooding directive "*Om vurdering og forvaltning av oversvømmelser (about*

Fig. 8.2 Permeable concrete surface on the ground outside a temple in Nara, Japan's first capital, about 20 min after some heavy precipitation

Fig. 8.3 Permeable concrete, Research demonstration, Kinki University, Kobe, Japan

Fig. 8.4 Permeable ground slab outside the Olympic stadium "the Birds Nest", Beijing, China

estimations and administration of flooding)" (2006/0005/ COD). In the document, analysis of flooding risks, the determination of what an acceptable flood risk level is and a demand for working out flood risk plans are requested. The reason for the directive is amongst others that flooding due to climate changes and development against more urban areas in Europe has increased considerably. Between 1998 and 2002, Europe experienced more than 100 larger destructive floods, up to 700 days in length and insurance losses of at least 25 billion Euro [3]. In May–June 2013 we experienced that the rivers Danube and Elben had floods with damages far bigger than any earlier registrations.

The Norwegian "Law about waterways and ground water" [4], Section 22.1.1 about *Natural flooding* is saying:

> With flood we mean that land areas are flooded with water. More precisely we describe floods in how much they exceed middle water level in the waterways. Normally it is described comparable to how often such a flood will happen, and it is normally described as a 10 year flood, 100 year flood and so on.
>
> It is first and foremost natural incidents as precipitation, temperature and snow melting that cause flooding. The water level in the water ways can vary considerably under normal conditions. In our climate it's normal with a spring flood due to snow melting, eventually in combination with precipitation. Direct rain floods normally happen in the summer or autumn, in the eastern part of the country. Winter flooding is most normal along the western coast, but has happened also other places in later years with mild winters.
>
> Ice can create special situations in the form of ice drift and ice dams that can cause collection of water and flooding when the ice breaks up.

The effects of larger floods can be formidable also in an erosion context.

In 2011 a record number of flood warnings were sent out in Norway: in total 69 warnings from 35 different situations.

Media information and report from many parts of the world tell about more often flood cases due to the climate change. Typically it is reported that the 20 years flood now comes every 3–5 years. The combination with increased urbanization makes the damages escalate. The question is whether we are able through political decisions and financial support to build out the needed precautions in pace with the recorded increase in damage risks.

NVE (Norway's Waterway and Electricity Directorate)'s flood warning is operative around the clock. The service has 12 hydrologist on watch constantly, checking data from 400 measuring stations around the whole country [5], so the problem is probably not that we are not warned in time but whether we have done necessary preparation beforehand.

In May–June 2013 there was large flooding both in Southern Norway and in greater parts of Middle Europe. German authorities warned about the largest damages ever in history. Pictures from Danube and Elben told about enormous areas under water.

The Norwegian Fund from nature damages claimed that the damages in the Gudbrand valley alone on buildings were about 240 million kroner (RMB) [6]. Norwegian Communities have obviously underestimated the effects and not yet taken the climate changes seriously enough in terms of planning.

The erosion ability of water depends on a number of factors, such as climate, topography, the property of the surfaces and vegetation. Roots of plants and vegetation absorb water, hold on the earth particles and reduce the ability of water to lead particles away. Areas will be exposed to deforesting much easier by opening to the negative erosion action of water.

Serious soil erosion has caused heavy damages to the national economic and social development and already becomes the most important environmental problem in China. Based on a report "the Water and Soil Erosion and the Control Measures in China" issued by Department of Water and Soil Conservation, Ministry of Water Resources of China, which goes as:

> In China, the types of water and soil erosion mainly include hydraulic erosion, wind erosion, freeze-thaw erosion, landslide, mud rock flow and hill avalanche, in which hydraulic erosion and wind erosion are most common and distributed most widely.
>
> ……Hydraulic erosion, …… mainly distributed in the vast areas south of the Great Wall and, in particular, is most serious in the middle reaches of the Yellow River and the upper reaches of the Yangtze River. In the Yangtze River Basin, soil erosion covers a total area of 562 000 km^2 and the induced soil loss is 2.4 billion tons annually; and in the Yellow River Basin it covers a total area of 450 000 km^2 and the induced soil loss is more than 2.2 billion tons annually. …… Lots of hill avalanche can be found in Guangdong, Hunan, Jiangxi and Fujian provinces and Guangxi Autonomous Region in South China; and lots of landslide and mud rock flow can be found in the upper reaches of the Yangtze River and the upper reaches of the Pearl River in Southwest China.

Chinese government has attached great importance to water and soil conservation and eco-environmental improvement and has made this a long-term basic national policy through implementing the National Plan of Water and Soil Conservation and Eco-environmental. China has gradually established relatively complete administrative system and legislation system for this purpose. Also through years of practices and demonstration projects starting from seven major drainage basins—the Yellow, Yangtze, Song-Liao, Hai, Huai and Pearl river basins and the Taihu Lake basin, a relatively complete set of rational national technical standards and measures have been formulated, issued and adopted for water and soil conservation.

The Plan defines the middle reaches of the Yellow River, the upper reaches of the Yangtze River, sandstorm areas and grassland areas as the key areas for water and soil conservation and eco-environmental improvement of the country. Strategies are provided to take key river basins as the key components, counties as units and small river basins as subunits, to carry out unified planning of hills, water, forest and road, and rationally to integrate structural measures, biological measures and tillage measures for moisture and soil preservation so as to realize comprehensive harnessing and development. Targets have also been made that initial effects will be achieved in 15 years with the soil erosion in

Fig. 8.5 It is said that water can move mountains. At least it is the cause for many strange formations. The picture is from the coast south of Chennai, Tamil Nadu, India

key river basins practically put under control, and that significant effects to be achieved in 30 years with the soil erosion in most areas practically put under control. The great objective of making all the territory eco-environmentally sound should be practically achieved by the mid-twenty-first century.

So far the practices for comprehensive control system and key engineering projects have shown fundamental progress in the harnessing of soil erosion and improvement of eco-environment as well as poverty alleviation. As the report summarized:

> In the Three Gorges Reservoir area in the upper Yangtze River key prevention and protection area, for example, after comprehensive harnessing of 7 years, the population carrying capacity has been increased by 6–23 persons per km². In Yulin Prefecture of Shaanxi Province, which is situated at the southern edge of Mawusu Desert, water and soil conservation has been implemented and farmland of 86,700 ha has been created with hydraulic filling and a lot of water resource projects have been constructed, thus fixing and semi-fixing 400,000 ha of floating sand out of the total of 573,000 ha and reversing the trend of desert encroachment.
>
> The works in the middle and upper reaches of the Yellow River reduce the sediment flowing into the river by more than 300 million tons annually.
>
> Poverty reduction as another example, in the upper reaches of the Yangtze River, the harnessing undertakings have eliminated poverty for more than 5 million farmers; in the key harnessing area in the upper reaches of the Yangtze River in Longnan area of Gansu Province, the coverage of poverty has been lowered from 66 to 24 %; and in the key harnessing areas of Sichuan Province, the percentage of poor households has been lowered from 15 to 5 %.

Some of the main types of water erosion are

- Rain and melting erosion
- River erosion
- Wave erosion
- Glacier erosion
- Ground water erosion
- Freeze-thaw erosion (Fig. 8.5)

Rain and Melting Erosion

If the ground stays without plant coverage for a while, the danger for erosion from rain and snow melting increases. The phenomena are well known in the agriculture, where people try to adjust the time for plowing and seeding to meet the runoff challenge. Heavy rain and large raindrops tear away soil particles that are led away with the runoff from the surface. Heavy rain might throw soil particles more than half a meter vertically and over a meter horizontally.

If the porosity in the soil masses is limited, for example through ice or frost, the eroded particles are led on to rivers and waterways.

The typical results might be

- Landslides in steep hills.
- Furrows on sloping ground
- Deep furrows in soft top roads without proper drainage (Fig. 8.6)

Fig. 8.6 Road between Yongsheng and Lijiang, Yunnan Province, China destroyed by landslide after heavy rain, September, 2004

River-Erosion

The waterways transport suspended particles from erosion in the river, surface water runoff and the industry. The suspended material might be small mineral particles, organic material, and other substances that keep themselves floating in the water. The amount of erosion depends on the geological conditions as well as the speed of the river, turbulence, and water volume. In addition to the suspended materials coming from typical bottom transport, some of the materials are sedimented or set off when the speed of the water in the river is reduced, for example in a bend when the topography flattens out or in the delta where the river floats into a lake or into the sea. Probably the most extreme example of river erosion and suspension of particles we find is in the Yellow River in China. (see also Chap. 5—Rivers and Lakes and Chap. 7—Transport Lanes) (Figs. 8.7, 8.8 and 8.9).

In a bending river, the speed of the water increases in the outer bend. The river will therefore erode most in the outer bend and deposit materials in the inner bend. These phenomena often lead to a situation where the river changes its shape and form a new channel on its way downwards. Flood can also lead to broken embankments and formation of new river channels and small lakes (Figs. 8.10, 8.11, 8.12 and 8.13).

Wave-Erosion

Wave erosion comes both from erosion from waves and water motion in harbors from ships digging in the bottom or at land, to continuous current and wave action along the coast, moving sand masses, changes beaches, makes small sea lakes, and stopping sailing routes, to waves that over years grinding rock and forming the strange formations along the coast (Figs. 8.14, 8.15, 8.16, 8.17, 8.18 and 8.19).

Erosion from a combination of currents and waves along beaches is a considerable problem for a number of coastal nations. Also wind might be an adding reason for the coastal erosion. Along the south western coast of Norway, there is beach area called Jæren. The changes of these beaches seem to be cyclic phenomena, where also wind is an important factor (Fig. 8.20).

At that time the Norwegian Minister of the Environment (from 2013 Foreign Minister) Børge Brende was proud when on December 12, 2003 he could tell about a new protection plan for the sand dunes at Jæren. This protection area includes 70 km long coastal strips.

Waves might also erode down rock formation through its continuous collision motion. The erosion increases when the waves contain sand and stone particles. The constant wearing can make the most peculiar formations on the rock formations Oceanside (See pictures from Koh Samui in

Fig. 8.7 Hukou (bottle mouth) Waterfalls, Shaanxi Province, China, the only waterfall in the course of Yellow River, where different erosions like eroded caves, hanging valley, cliff waterfall, flow erosion, inclined bedding, etc., are evidenced in this geo-park. September 2008

Fig. 8.8 Entrance of Yellow River to Bohai Sea, Dongying, Shandong Province, showing a clear soil erosion by the Yellow River. October 2008

Chap. 10—Recreation). This has been explained by ancient Chinese idiom originated from Song Dynasty which means that constant dropping will wear away a stone.

The ability of water to bring large quantities of sand and other materials over a rather short span of time is formidable.

The tropical hurricane Katrina was formed over the south eastern Bahamas on August 23, 2005. First it was somewhat reduced in strength as it crossed over Florida, USA as a category 1 hurricane. In the Mexican Gulf, Katrina picked up strength and was upgraded to a category 5 hurricane.

Fig. 8.9 Another Yellow River in Darkhan, Mongolia, as told by Prof. M.R. Hansen, School of Mining and Materials, University of South Dakoda, USA who had been voluntarily to Mongolia with her wife Barbara for ten years teaching Concrete and English. Though incomparable to China's, the eroded river winding in the grasslands still aroused interest of author to take a close view. June, 2014. *Photo* Bin Wang

Fig. 8.10 Erosion and erosion protection, Moscow River, Russia

When it swept over the utter part of the Yucatan peninsula in Mexico, the material damages were formidable. All the leaves were blown off the trees in a belt of about 10 km. Hotels at the beach in Puerto del Carmen and at the island of Cozumel during the restoration work had to get rid of about 1 m of sand from rooms and reception areas in the ground floor.

Even better known was probably the damages in New Orleans and Louisiana where the number of deaths was large. Katrina was afterwards recorded as the largest natural catastrophe in modern American history until then. Waves were throwing houses and cars up to 10–20 km inland in Louisiana.

Waves are found in a number of formations, heights and speed, from the smallest waves of a few centimeters to the recorded waves of up to 100 m high created by continuous action of wind over large oceanic distances. The waves that are created from tsunamis are normally not so high in the ocean, but when they reach land, speed, bottom conditions, and the topography on land will create wave heights and abilities far into and high over the coastal zone.

We build wave breakers and sometimes large breakwater peninsulas to protect our harbors (Fig. 8.21).

Glacier Erosion

The glaciers are a combination of snow and ice, but they are also in constant "slow" motion. Underneath the glaciers the erosion takes place, from wearing and picking.

Glacier Erosion

Fig. 8.11 Concrete slabs to protect an old church in Uglich against the forces from the Russian mother river Volga

Picking is when the glacier forms its movement it picks up rocks and other loose material. The loose material is normally formed when water in the bottom of the glacier freezes and creates frost stresses. The loose material freezes into the glacier which brings it along.

Wearing happens when the glacier carries loose sand and rock and moves slowly over the rock like a giant sandpaper. Some of the wearing materials also contribute to wearing down the bedrock from the bottom (Fig. 8.22).

Ground Water Erosion

Ground water can dissolve and wear away certain minerals in the rocks and create grottoes, underground channels, and waterways. When the water level in such underground channels sinks, the stabilizing pressure against the roof is

Fig. 8.12 The rivers also bring with it a lot of organic material. The picture is from Chao Phraya-river in February, just up from Bangkok, Thailand,—many months after the end of the rainy season

Fig. 8.13 Also the animal and bird life is affected by the erosion

Fig. 8.14 Sea wave-eroded cliffs now become nice perching place of seagulls. Changdao (Long Island), Yantai, Shandong Province, China. *Photo* Fuxin Yu, June, 2014

reduced, and the possibility for materials to loosen from the roof increases. In the end, this might lead to that the grotto opens up all the way to the daylight.

Such underground formations and structures are found all over the world, but possibly the most well know of all of them are the so-called *cenotes* on the Yucatán—peninsula in

Fig. 8.15 Xianren (Immortal) Bridge, Rushan, Shandong Province, formed by sea water erosion. July, 2014. *Photo* Xianglai Yu

Fig. 8.16 Rock of Aphrodite, the birth place of God of Love, where different types of water erosion can be found. Cyprus, end of March, 2012

Fig. 8.17 A different type of "EROSION" caused by impolite visitors are evidenced by the tree in the roadside near the Rock of Aphrodite. Cyprus, March 2012

Fig. 8.18 Concrete "blanket" as erosion protection, Color Line's ferry terminal, Strømstad, Sweden

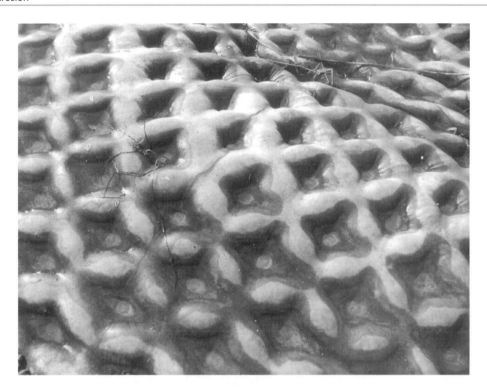

Fig. 8.19 Concrete "blanket" as erosion protection, Color Line's ferry terminal, Strømstad, Sweden

Fig. 8.20 The grass is important as protection of the sand dunes at Jæren

Fig. 8.21 Sailing past the opening in the long molo to the harbor Fredrikshavn, Denmark. The protection breakwater is covered with large rocks for erosion protection. The harbor is the foundation for the trade and industry in Fredrikshavn, which annually has about 5 million ferry tourists going through. The harbor is also important for the local fishing industry and one of the most important naval bases in Denmark. It is one of the 10 busiest harbors in Denmark

Fig. 8.22 Platform of 4506 m above the sea level, Yulong (Jade Dragon) Snow Mountain, Glacier Geo-Park in Lijiang, Yunnan Province, Southwest China, where glacier formation is "pouring" down along the valley like an ice river. August, 2013

Fig. 8.23 Cenote, Margaret River area, Western Australia

Mexico. In the north and northwest part of the peninsula, the underground water channels might go down to 50–100 m below surface. Cenotes is a Maya Indian expression, but this expression is used on similar phenomena also in other countries and continents. It is claimed that some cenote-channels might be up to 100 km long.

The most famous Maya cities from ancient times, as for example Chichen Itza, are all located with such underground water sources. The Maya's regarded their cenotes as holy, and they believed that this was the entrance to the underworld.

The water in the cenotes is normally very clear after the rainwater having been filtered through many meters of rock. Diving or swimming in the cenotes is a very special experience. The rather limited animal life has very little pigment, as for example numerous, completely small white crabs.

The Yucatán—peninsula is relatively flat, and in some areas the fresh rainwater mixes with sea water (Figs. 8.23 and 8.24).

Fig. 8.24 Nice view of limestone caves, where erosion and formation coexist, the stagmalites grow from opposite direction and eventually meet over time. Ludi Rock, Guilin, Guangxi Zhuang Autonomous Region, China. September 2009

Fig. 8.25 Rock cut protected by sprayed concrete, Old Drammen Road, Sandvika, Norway

Freeze-Thaw Erosion

When water freezes in cracks and fissures in rock and stones, the ice will produce stresses across the crack, which leads to new crack formation. This again leads to frost deformation and over time to the erosion. An important impact of freeze-thaw erosion might be desertification followed by degradation of vegetation, soil degradation, the exposed and fragmented surface, etc. This already happens in China in the Qinghai Tibet Plateau Permafrost Region of 160×10^4 km^2. This normally occurs in a degree of excessive development from sporadic distribution, zonal distribution and further to the patchy distribution.

In particular, free-thaw erosion is a challenge in cutting through the topography that is constructed from blasting, etc. A normal remedy against this is the use of sprayed concrete to protect the rock (Figs. 8.25 and 8.26).

In areas with frost in late winter or early spring and just as sure as the birds of passage coming, we find notes in the media about road damage on the roads. The frightening thing is that this is observed not only on old roads, but also on the new ones ready only a few years ago. The Norwegian Road Authorities from time to time established a new expert group, which seems to come up with the same recommendations as the last group:

- Prepare for adequate drainage of the water.
- Prepare the top road base thick enough in comparison with the expected frost and the ground conditions.

Fig. 8.26 A special type of freeze-thaw erosion is ground frost rising. There are many places, where you can observe this after the damage of the road surface. Fine materials in the road base observe water and these freezes in the winter. The results are easy to observe when the spring comes

- Make sure that there are not too much fines in the road base.
- Make sure that the stone/rock size in the road base is not too large compared to the layer thickness.

The road authorities have handbooks telling how to design properly. They arrange seminars and conferences to ensure that the proper knowledge is brought forward to the right persons. Still, every year we have to spend new

Fig. 8.27 Not an unusual spring message from the Road Department (Teleskade = Frozen ground damage)

millions to repair for frost damage. The solution—probably in the intersection between better understanding of the erosion potential of water and even more better education and much stronger economic responsibility (Figs. 8.27 and 8.28).

Also Human Created Structures Are Attacked

On February 12, 2013, the Norwegian Government Road Department arranged a seminar to present analysis from their new development program *Lasting Structures*: The program is a development drive for improvement of bridges and tunnels. Many of the lectures at the seminar told about the efforts and the challenges to find better solutions to reduce the attack from the always present, simple, neutral attacker, water.

A Swedish guest speaker said the following from experiences in Sweden: The most heavy attacking environmental factor is water: Water that freezes, water that expands, water that dilutes chlorides and other chemical substances and attacks structures, water that stays long and increases the attacking time, water that spreads and enlarges the attacking area, water that weakens materials. Water as a base for biological attack, water that short-circuit functions [7].

The most important building materials include cement, lime, and gypsum, all of which needs water to become binders for buildings and form concrete, hydrated lime, and

Fig. 8.28 Combinations of water and human activity can be an efficient erosion mechanism

plaster, respectively, through hydration reaction with water. The deterioration of these building materials on the other hand can also be attributed to water directly and indirectly.

References

1. Access on March 9, 2014. http://wenku.baidu.com/view/3c2256160b4e767f5acfce75.html.
2. Access on March 9, 2014. http://en.wikipedia.org/wiki/Coriolis_Effect.
3. Lidholm Oddvar: *Overflatevann – Utfordringer og muligheter (Surface Water – Challenges and possibilities)*. Norsk Betongforenings Rapport nr. 3 (Norwegian Concrete Society, Environmental Committee Report No 3) – *Permeabel betong og overflatevann (Permeable concrete and surface water)*, Editor Per Jahren, Oslo, Norway, 10.10.2011.
4. Olje – og Energidepartementer (The Oil – and Energy Depaertment): *Lov om vassdrag og grunnvann (Law about waterways and ground water)*. NOU 1994: 12, Oslo, Norway.
5. NVE: *Store flommer i Norge (Large floods in Norway)*, Annonsebilag Aftenposten, Oslo, Norway, October 2012.
6. Per – Ivar Nikolaisen, . *Stadig dyrere flommer (More costly floods)*. Teknisk Ukeblad, Oslo, Norway, 19/20 2013.
7. Freiholtz Bernt og Helsing Elisabeth: *Bestendiga konstruktioner – Livslengds- och bestendighetsaspekter på de svenska tunnlarns och broarna*, Seminar Varige Konstruksjoner, Statens Vegvesen, Oslo, Norway 12. February 2012.

Water Consumption

Even if the Romans were not the first to have an effective water supply network, it is impossible not to be impressed over what giant technological and healthwise step they did with their water supply technology, 2000 years ago.

Water over time became a very important issue for the Romans—in more than one way. Even in comparison with modern terminology and measures, the Roman "water market" became considerable. In a book from 1965 [1] Hjalmar T. Larsen has written about the history of bathing. From the book we have found interesting data about the bathing in Rome 2000 years ago.

In the beginning, the Romans were scorning the Greeks for their large baths. Only under Cato the older and in the time of Scipio Africanus, the Romans started bathing the Greek way. The Adile Marcus Vipsanius Agrippa, in his time built 170 saunas in Rome, and built the first termes in the Greek fashion (19 BC.).

These baths got very popular, and the Roman rulers saw the advantage in buying the popularity of the people in the building of termes, one more advanced and bigger than the other. Emperor Nero built his terme in the year 64, called "The Golden House", about which said: "Who is worse than Nero, and what is more beautiful than Nero's bath…". Later many emperors built termes: Vespian (year 68), Titus (year 75), Trajan (year 103), Hadrian (year 120), Caraculla (year 217), Diokletian (year 295) and Constantin (year 324). At Constantin's time, Rome had 900 public baths and 11 termes in operation (Fig. 9.1).

The size of these structures was impressive. The Titus-terme had over 100 bathrooms. Caracallas' terme had a surface area of 124 000 m^2 and a water area of 1300 m^2. For the 2300 peoples that could use the bath at the same time, there were 1600 polished marble chairs.

Diokletian's terme was a bit larger, with 125 000 m^2 water area and 1700 m^2 water area. 3600 people could bath there at the same time (Figs. 9.2 and 9.3).

At Emperor Constantin's time, Rome had a water consumption of 750 million liters per day [1]. Other sources mention higher numbers. Wikipedia claims 1 million m^3 per day [2]. It is claimed that this is 26 % more than the water capacity of Bangalore in India, serving 6 million people.

Rome had a well-organized water and sanitary system. As early as 350–312 BC., Appius Claudius Caecus built the first water supply channel—Aqua Appia. 40 years later came Anio that could carry twice as much water as in Appia.

Aqua Appia led good river water from far away, and in the end there were 11–12 such water channels supplying Rome with one million inhabitants with water. Some of these channels also gave us the fantastic water bridges, the aqueducts—which many first and foremost think of when the Roman water systems are mentioned.

The Romans also brought with them their water building art to other parts of the Roman empire, where hundreds of water building structures were built (Fig. 9.4).

Along the River Rhine, the Romans had much activity for 200 years from about year 70 to year 276. For the city of Cologne was built a 105 km long water line, supplying the city with 30 million liters of water every day from Eifel. A major part of this 90 km long water culvert was built in years 70–80 Roman type concrete. As a binder they used the pozzolan "trass", and a cement made by burning of quartz and clay. The water culvert was in use for 450 years. Then it was demolished and cut into building blocks used in city walls and fortifications.

Probably the most well-known of the Roman aqueducts is—Pont de Gard in France.

The Roman engineer and architect tell in his famous book that the gradient in the aqueducts not should be less than 1–4800. The gradient in Pont de Gard is recorded to 34 cm per km., which is somewhat "steeper" than 1–3000. The total fall is 17 m in 50 km. The Pont de Gard aqueduct could supply 20 000 m^3 per day [3]. In addition to the water structure technology, the building of these structures also meant fairly advanced technology in surveying and maintenance (Fig. 9.5).

The best kept aqueduct on the Iberian peninsula, and one of the most remarkable of such structures is found in the city of Segovia in Spain, a mountain town about 100 km north

Fig. 9.1 Caracallas terme, drawing from 1899

Fig. 9.2 The ruin from Caracallas Terme is a popular sightseeing object in Rome. The building of Caracalla's bath was started by Septimus Severus in 206, and opened by Caracalla in 217. There were on the inside clad with granite, basalt and alabaster

Fig. 9.3 Remains of Public Baths, Kourion, Cyprus, which can be dated back to ca 50 B.C.—100 A.D. Kourion city in the ancient time has experienced through different phases spanning the Hellenistic, Roman, and Christian periods, the city therefore not only has a very large Agora (market place) and an early Christian Basilica within the city walls, but also has large public baths equipped with cold (frigidarium), warm (tepidarium) and hot spas (caldrium) or "Thermae" as social places where people at that time gathered and discussed. March 2012

west of Madrid. Segovia is located on a cliff formation high above the local river. The Romans therefore choose to lead the water to Segovia from Fuente Fria—a river up in the mountain, 15 km from the town. First the water was collected in two water tanks, and decanted in sand, before it is lead 728 m with 1 % fall on the aqueduct.

At its highest, this structure is 28.5 m high. It is constructed by local granite blocks without mortar, and has altogether 167 arches. There is no concrete in the arch structure, and somewhat uncertain if it is in the channel itself.

However, the Romans were far from the first that performed water building wonders. An example is the nomad people Nabateer's, that 300 BC used cement mortar to clad and tighten their wells in the desert [4]. In this way, they secured their traveling routes through the desert, and became a leading trading community. Their special secret technology made it possible for them to travel further in the desert than other nomadic people. It is also claimed that the Phoenicians used concrete to clad wells 500 BC, and that the Greek did the same.

There exist a number of calculations about how much water that modern people as a minimum needs every day. As approximate figures we can use:

	Daily	Daily over time
Drinking water	3–4 L	3–4 L
Cooking	2–3 L	2–3 L
Personal hygiene		6–7 L
Cloths washing		4–6 L

These minimum numbers are of course far below the normal consumption in industrialized countries.

To be sure of a reasonable life quality standard, it is regarded that each person needs a bit less than 100 L per day. This means 30–40 m^3 per person per year.

Fig. 9.4 There are still ruins of the aqueducts in Rome, but unfortunately not so many of them are still there. The picture is from the ruins of the Claudius—aqueduct

Fig. 9.5 Roman aqueduct in Segovia, Spain

In the UN centennial goal they have quantified the need for water this way [24]:

> It is a human right for each individual to get about 20 L of water every day, independent of wealth, localization, sex, or racial background, ethnic background or other group differences.

The difference in water consumption varies considerably from one part of the world to the other. For example—60 % of the world population lives in Asia and the Pacific, but they represents only 36 % of the water consumption [5]. The consumption per person in USA is over 500 L per day, while

9 Water Consumption

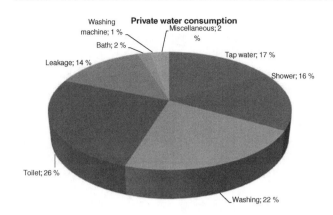

Fig. 9.6 The spread of the private water consumption in North America [5]

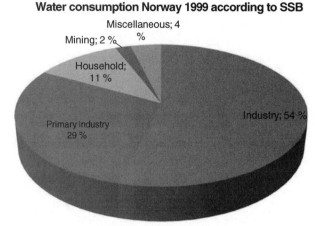

Fig. 9.8 The water consumption in Norway [16]

it on Madagascar is well below 10 L. (We remind about the historical fact that the water capacity per person in Rome, nearly 2000 years ago, was about the double of what is consumed in USA today.) (Figs. 9.6, 9.7 and 9.8)

The large proportion of the water consumption that goes for agriculture has its challenges. This is particularly important for developing countries where water for agriculture normally takes more than 60 % of the total water consumption while it is well below 50 % for developed countries. A special case for Canada gives a very low share of less than 20 % for agricultural water use. It is believed that the world population will increase within the order 2–3 billion people in the next 40 years. Estimates say that this, in addition to the general increase in wealth or living standard, will increase the food demand with 70 % in the same time frame [5]. Consequently—this will also increase the water demand considerably. No wonder The United Nations World Water Development Report claims that "There is a strong positive link between investment in irrigation, poverty alleviation and food security." [6] (Fig. 9.9).

More efficient use of water in agriculture is therefore a very important goal in many countries where this is necessary. However—this gives some strange side—effects:

Saudi Arabia—one of the largest grain producers in the Middle East, has decided to reduce their production by 12 %, to lessen the burden on non-sustainable use of ground water. Instead the country will rent areas in Africa for a similar production. Likewise, India is growing corn, sugar canes and lintels in;—Ethiopia, Kenya, Madagascar, Senegal

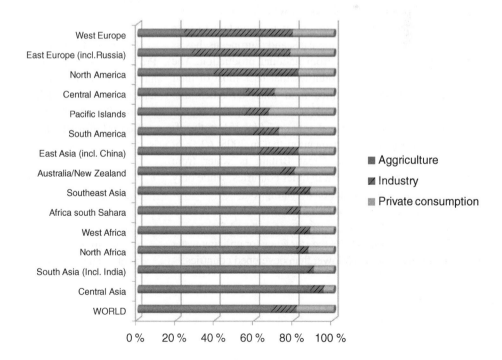

Fig. 9.7 There are considerable differences in how we use the water that we have available [5]

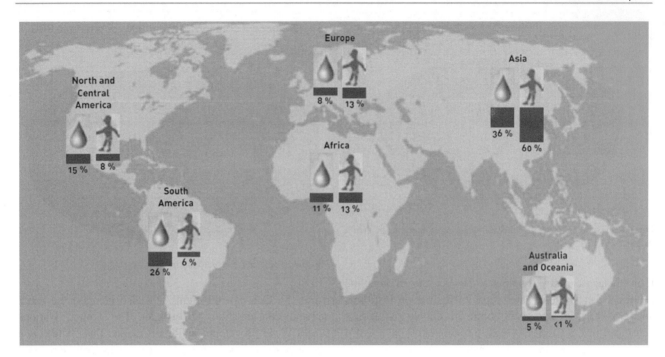

Fig. 9.9 The available freshwater distributed regionally versus population. The global overview of water availability versus the population stresses the continental disparities, and in particular the pressure put on the Asian continent, which supports more than half the world's population with only 36 % of the world's water resources [6]

and Mozambique, to cover the food demand in India, without at the same time increase the Indian water consumption [5].

On the other hand, the water consumption for food production in Western Europe and North America has been considerably reduced in the latest 10 years.

It is claimed the reason for the high North American water consumption is the relatively low price on water.

In a water consumption context, it is important to note that the footprint of the Americans and Europeans has is greater than what they gets from national production. The import of food from other countries that uses most of their water consumption to produce food must be taken into account. As an example, the UN water development report: *Managing Water under Uncertainty and Risk* [5], with reference to Chapagain and Orr mention that in 2008, 62 % of the "water footprint" from Great Britain is water from agricultural products produced in other countries, while only 38 % originates from the nation consumption.

A report from Ministry of Water Resources of China indicates that in 2008 besides the larger percentage of water for agriculture of more than 60 %, the efficiency for utilizing water for agriculture is also much lower than developed countries:

Country	Grain production per m^3 of irrigation water (kg/m^3)	Effective coefficient of irrigation water utilization	Ratio of recycling of industrial water use (%)
China	1.0	0.48	62 %
Developed countries	2.5–3.0	0.7–0.8	80 %

The State Council of China approves a Comprehensive Programing of National Water Resources, in which targets have been designated as follows in view of the present situation and challenges including the increasing population up to 1.6 billion by the year 2030:

Control of total water consumption within 700 billion m^3 is shown in the Fig. 9.10.

- Continuous efficiency improvement: by control of the total water consumption within 670 and 700 billion m^3 in 2020 and 2030 respectively, the water consumption intensity per unit of GDP will be reduced by 50 % and further by 40 % on the basis of 2008 and 2020, respectively. The effective coefficient of water for agricultural

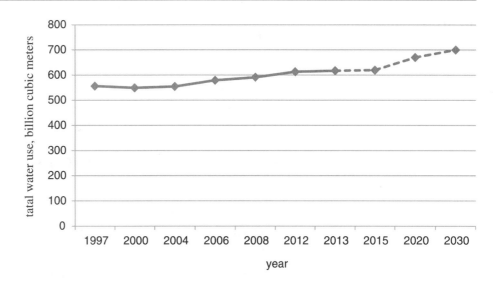

Fig. 9.10 China total water consumption and perspective

irrigation will be increased to 0.6, a big improvement of 0.48 in 2008.
- The water consumption will be re-structured to meet the demand of growing urbanization and industrialization.

Year	Agriculture (%)	Industry (%)	Urban (%)
2008	62.0	23.6	14.4
2030	59.8	25.2	15.0

Among which the structure of water allocation for urban use and rural use will be changed from the present ratio of 31:69–37:63 in 2030.

The large consumption of water for agriculture has important negative by effects. The irrigation combined with bad planning of runoff from precipitation, leads to un-necessary pollution from manure storage, silos and plant chemicals. The result has both been found in drinking water sources, fish deaths, and increased algae growth in the water ways. As mentioned in Chap. 4 about the oceans, this might also be an important indirect cause for coral deaths.

The water consumption varies considerably with the type of agricultural product that is produced. An UN—related report [5] for example claims that 15 m^3 of equivalent water are needed for each kg of beef that is produced. More details are given in the table below (Fig. 9.11).

This table gives examples of water required per unit of major food products, including livestock, which consume the most water per unit. Cereals, oil crops and pulses, roots and tubers consume far less water.

10 % of the population in the world lives in dry areas with a precipitation of less than 300 mm per year.

Even the driest part of the world has enough access to cover the need for a private consumption of about 100 L per person per day in a reasonable way. The question is however how we distribute and administer the water supply. For example, it is estimated that the Amazonas River has a flow of water similar to about 30 times the total primary need in the world.

On 1, November 2011, we celebrated the birth of a girl child in the Philippines being the citizen of the world number 7 billion, where four and half billion lives in Asia and Africa. In 1970 there were 3.8 billion people in the world, with more than 2 billion in Asia. The growth in the population, water consumption and lack of water are closely related. Demographers have for example estimated a strong connection between the education of girls and the population growth. They claim that in average each female without education will have 4.5 children, female with a few years of basic education will have 3—and women with at least 2 years of education after basic school will in average get 1.9 children [7].

The population growth in the world has been drastically reduced in later years, but there is still a long way to go before we have a reasonable balance. Some estimates say that we might get a balance towards the end of this millennium. This means that the population growth and social development will increase the pressure on the water resources in world in many years ahead (Fig. 9.12).

Desalination

Desalination of salt water for use as drinking water has got considerably increased interest in later years, and there is an important growth in this sector. A number of new technologies are under testing and development. The most used methods are distillation or use of a membrane filter/opposite osmosis. Desalination is regularly used in ships and submarines. Wikipedia [8], with reference to International Desalination Association, told that in 2009 there were 14 451 desalination plants in the world in which more than

Fig. 9.11 Water requirement equivalent of main food production

Product	Unit	Equivalent water in cubic metres
Bovine, cattle	head	4,000
Sheeps and goats	head	500
Meat bovine fresh	kilogram	15
Meat sheep fresh	kilogram	10
Meat poultry fresh	kilogram	6
Cereals	kilogram	1.5
Citrus fruit	kilogram	1
Palm oil	kilogram	2
Pulses, roots and tubers	kilogram	1

Source: FAO, 1997b.

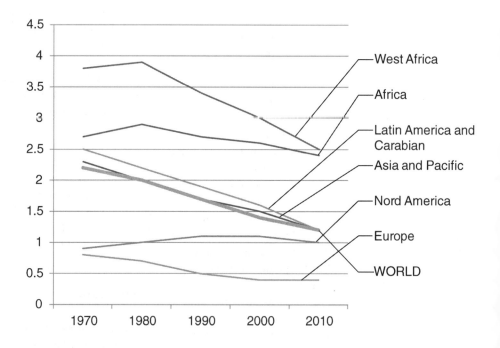

Fig. 9.12 The population growth in the world in % [7]

150 countries and regions have been engaged. These plants produced nearly 60 million m³ per day, meaning an annual growth of 12.3 %. It is expected that this number will increase to 120 million m³ in 2020, where 1/3 is planned in the Middle East.

So far the largest seawater desalting plant is located in Saudi Arabia with the first phase project with a capacity of fresh water production more than 1 million m³ per day and power generation capacity of 2600 MW commissioning in April 2014. Before this Haifa in north Israel has the world largest desalting plant with a capacity of 127 million m^3 per year accounting for half of that of the total in Israel.

China started the production through seawater desalting in 1981 in Xisha Island, Hainan Province. Based on a report from China Environment Membrane Separation Engineering Center, by August 2011 China has built 66 desalting plants (including 6 plants already shut down) producing 756.8 thousand m^3 of fresh water per day. Shandong takes more than half of the total seawater desalting for fresh water capacity in China, among which Qingdao, coast city of east of Shandong Province has built a total capacity of 2 million m^3 of fresh water production.

Desalination has a number of challenges;

- The cost—these are expected to be somewhat reduced as the technology is further developed, and the plants increase in size. Costs in the order ½ USD per m^3 are mentioned by several sources.
- Utilization/depositing of salt/or salty liquid can be a problem unless there is there are sufficient alternatives.
- Energy consumption and CO_2 mission from the energy source. This is in particular important in the case of distillation technology.
- Acceptable water intake that avoid negative effect on the destruction of life in the sea as plankton, fish eggs and larvae.
- The desalination process has great requirements to the materials in the structures due to possible corrosion.

In some parts of the world, desalination structures have been built in combination with other industrial activity as a form of circular industry chain to reduce possible negative environmental factors.

Lack of Water

Safe access to drinking water is absolutely essential for the life on earth—including human beings. Over the last hundred years this has been improved for most people in the world. 1.6 billion people have got access to safe drinking water since 1990, and the part of the population in developing countries that has got access to good drinking water has nearly increased linearly from 30 % in 1970 to 84 % in 2004.

WHO, UNICEF reported in 2010, that since 1990—510 million people in East Asia, 137 million in South Asia and 115 million in South East Asia have got access to tap water at home [5].

One of the millennium goals was to halve the number of people that did not have access to safe drinking water and sanitary installations before 2015. This goal is expected to be achieved. A UN Report in 2006 claims that there is enough water for all people in the world, but the access is obstructed by bad leadership and corruption [9].

In an article in China Daily 27. May 2013, Noeleen Heyzer, Under General Secretary in UN and Executive Secretary in UN, as well as Executive Secretary of ESCAP (UN Economic and Social Commission for Asia and the Pacific), claims [10] that more than 2 billion people in the world got access to safe drinking water in the years 1990 to 2010. This means that the millennium goal was met in 2010 —5 years ahead of time. However, it is still so that more than 600 million people will lack safe drinking water in 2015, and 1.7 billion people in Asia and the Pacific are without acceptable sanitary water conditions. In that area, there is still a long way to meet the millennium goal.

Water shortage is still a great challenge when it is reported that nearly 2 million children die every year due to unsafe water access and unacceptable sanitary conditions, and that there still are some 2.6 billion people that do not have these commodities. In China, the world most populated country and one of the 13 countries least lack of fresh water, there exists a huge challenge for water shortage—with only 1/4 of the world average fresh water resource per capita. The uneven distribution of fresh water upon seasons and geological locations makes the problem more complicated and serious. This is a formidable and impressive development for people's living conditions, the water resources is influenced by lack of water in some provinces and water pollution caused by a combination of; increased centralization, rapid industrialization and increased economic growth not followed up by environmental knowledge. Some observers have made estimates saying that half of population in the world might have some kind of water shortage in 2025, and even USA might experience water shortage in a few years.

To take care of our water resources is definitely a global problem, while lack of water is a local, and sometimes a regional challenge. There is enough water in the world, but not always locally available with sufficient quality, and at all times of the year. Even if it most often is a local problem, there should be no doubt that this should be handled as a global challenge.

During a moving national seminar to 4 cities in India in 2011, the President of Indian Concrete Industry, Vijay Kulkarni, said in a lecture [11], the following regarding the water situation in India; *India is the second country the world having highest precipitation. However, due to increasing population and pollution due to human activity, the supply of water is reducing. Further, rainfall is not uniformly distributed and erratic. Total annual precipitation is about 4000 km³ and an average of the rainfall received is 1200 mm. There are annual rainfalls of 11 000 mm in Cherrapunji and the minimum average rainfall in West Rajasthan of about 250–300 mm. As per the World Watch Institute, India will be a highly water stressed country from*

Fig. 9.13 Rice harvesting is going on in Tamil Nadu, India. Rice is the largest food commodity in the world, and half of all rice in the world is grown and consumed in India and China. Large quantities of water in the planting season are needed. In the largest production countries, the monsoon rain is the most important bases for the growth

2020 onwards. The meaning of water stress is that less than 1000 m³ of water will be available per person per annum. India's population, which is projected to go up to 1333 million by AD 2025 and further to 1640 million by AD 2050. It is projected that the per capita water availability in India may reduce to about 1200 m³/year by 2047 (Figs. 9.13 and 9.14).

Many areas in India are perpetually drought-prone. A recent study, which identifies such areas, concludes that in most parts of India the probability of moderate drought ranges from 11 % to 20 %. Water is required for concrete production and curing in large quantity. The availability of water may therefore be one of the major constraints to growth of the concrete industry in perpetually drought-prone areas.

Another well-known concrete technologist and "guru", Professor V.M. Malhotra from Canada gave the following alarming data in a lecture in 2009 [12];

> The IPCC reports on climate change issued in 2007 warn that global warming will affect very seriously the availability of water in the future. The Himalayan glaziers are melting fast. This could lead to water shortage for hundreds of millions of people. The glaziers regulate the water supply to Ganges, Indus, Brahmaputra, Mekong, Thanklwin, Yangtze and Yellow Rivers are believing to be retreating at a rate of about 10–15 m every year. It is estimated that 500 million people on the planet live in countries critically short of water, and by 2025, the above number will leap to 3 billion.
>
> *... USA is the leader in water usage as shown below:*

Continent	Water Consumption in liters per day
North America	600
Europe	300
Africa	30

In spite of the looming water crisis in the not too distant future, there is a huge wastage of water worldwide. For example, 9.5 billion liters of water it would take to support 4.76 billion people

Fig. 9.14 Harvesting goes on in the dry season. A blacktop asphalt road can be useful for the hand handling. The traffic has to wait. From Tamil Nadu, India. End of February

of their daily needs as set by the United Nations. On the other hand, currently 9.5 billion liters of water are being used to irrigate the world's golf courses.

> ...The intense irrigation has dramatic effect on the water tables. For example, the number of bore holes that pump irrigation water to India's farmland was 10 *000 in 1960. And the number increased to 20 000 000 in 2007. This has caused declines of water tables from 100 to 150 m in some places.*

In addition, we again remind about the fact that about 70 % of all fresh water consumption goes for use in agriculture, 20 % goes for the industry, while the private consumption takes 10 %. The numbers varies considerably from one part of the world to the other, and in some typical industrial countries, the industry is the biggest consumer. In Belgium for example, the industry takes 80 % of the water consumption [13].

Chinese statistics [14], tells about a very strong effort in letting the agricultural population getting access to clean water from the tap, that have led to an increased direct access to water from the tap at home for rural population from 61.3 % in 2005 to 72.1 % in 2010, a big improvement compared with the numbers of 30.7 % in 1990 and 58.2 % in 2003. The target is 100 % direct access to tap water for rural population by the year 2015.

The water consumption will to some extent relate to the price of water, Many, probably too many, do not think that they have to be careful with the water consumption, because it does not cost so much. The water and sewage prices in Norwegian communities, however, varies quite a bit, from under 4000 kroner (RMB) to nearly 20 000 kroner (RMB) per year based on a expected water consumption 150 m^3, meaning from under 30 kroner (RMB) to 120 kr (RMB) per m^3 (2012).

The water prices have increased considerably in later years. For example, The city of Drammen tells [15] that the water consumption is stipulated to 110 m^3 for a unit below 65 m^2–280 m^3 for living unit above 160 m^2. The prices for water per m^3, has increased from 12.53 kroner (RMB) in 2010 to 13.93 kroner in 2011 and 15.49 kroner in 2012.

According to the Norwegian Statistical Central Bureau (SSB) [16], the financial cover percentage for the water works in Norway as whole is 101 %. This means that the net result is 1 % higher than the total income from what the inhabitants pay plus the capital costs. There are large differences throughout the country (2003), where the county of Telemark has the highest cover percentage with 112 %, and the county of Nordland the lowest, with 91 %.

In total, we had 1544 in Norway in 2003, where 622 got water from lakes, 379 got water from rivers and brooks and 574 from ground water. About 90 % of the Norwegian population gets their water from surface sources.

If we look at some different industrial countries that might be comparable, the prices for water vary considerably (Fig. 9.15).

The price of water also varies from place to place, from time to time and from sector to sector in one country. Based on Chinese government guideline for water pricing, progressive pricing policy has been applied, and normally three categories of household water consumption are classified for ascending prices. The basic water price for city residents in Shanghai is about 85 % of that of Beijing. However, the classification on water consumption is different. The rapid increase of household water price in recent years in Beijing may demonstrate the increasing pressure on water shortage in the capital and the economic solution of water conservation.

Water price at basic level in recent years, Beijing RMB/m^3.

Item	2012	2013	2014.5
Tap water price	1.70	1.70	2.07
Water resource cost	1.10	1.26	1.57
Sewage treatment cost	0.90	1.04	1.36
Overall Price	**3.70**	**4.00**	**5.00**

Water Pricing being effective from May 1, 2014, Beijing.

Water supply	Level	Yearly household water consumption m^3	Total price RMB/m^3	RMB/m^3		
				Water cost	Resource fee	Sewage treatment fee
Tap water	Basic	0–180	5	2.07	1.57	1.36
	Middle	181–260	7	4.07		
	High	>260	9	6.07		
Well	Basic	0–180	5	1.03	2.61	1.36
	Middle	181–260	7	3.03		
	High	>260	9	5.03		

Non-household water consumption:

- Schools, Social welfare organizations, etc.: unified price of 6.0 RMB/m^3, progressive pricing effective
- Restaurants, industrial and commercial use: 7.15 RMB/m^3 from May 1, 2014 and 8.15 RMB/m^3 from January 1, 2015;
- Special sectors such as car washing, bath industry, golf courts: 160 RMB/m^3.

Studying the prices of water in the world clearly indicates that the prices in no way mirror the access to water. The consumption of water per person per year, however, varies

Fig. 9.15 Prices for water 2001 [17]

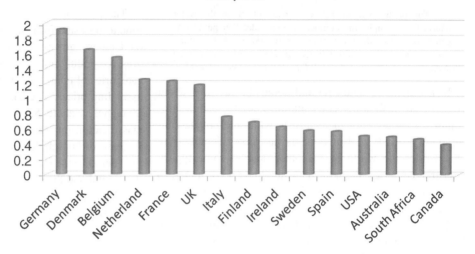

strongly, from for example (2004); Canada—1 420 m³, Germany—430 m³, Belgium 650 m³ [17]. It is also interesting to note that drinking water in bottles costs 240–10 000 times more than the water from the tap [17] (with reference to United States Natural Resources Defense Council). The bottle water price is based on the funny effects in the market place, and has nothing to do with the production price (Fig. 9.16).

It costs to "produce" water, and the use of water has variable effects on the economy of people. Statistics from El Salvador for example, indicates that the 20 % poorest part of the population must use 10 % of their income on water. In UK, the authorities has defined the limit for use of the income for water to 3 % to make sure that water must not be a burden to people [18].

Fig. 9.16 At the street in Shanghai a half liter bottle costs 2 RMB, when the street salesman has a refrigerator—if not, the price is 1 RMB. At a hotel—the price might often be RMB. Similarly we can buy a bottle of mineral water for 1 RMB while this might cost 10 RMB when it comes to the top of the Mt. Tai, one of the top four mountains in China

A UN report gives the following approximate numbers for what part of the food budget that goes for water supply in some poorer countries in South America [17]:

Guatemala	3 %
Peru	4 %
Paraguay	6 %
Mexico	4 %
Suriname	8 %
Columbia	8 %
Bolivia	8 %
Nicaragua	9 %
Ecuador	9 %
El Salvador	11 %
Argentina	11 %
Jamaica	11 %

The mentioned UN-report also states that in general, poor people pay more for water, both in relative value and absolute money value. This is correct both in a comparison between poor and rich countries and within each country. This obvious injustice is one of the greatest injustices with respect to human rights, the right to clean water is one of the greatest social problems in the world. Nothing, possibly with the exemption of the air we need to breath, should be more a common property than water (Fig. 9.17)!

The Norwegian water consumption per capita is a bit higher than the one in France.

Some health institutions claim that we each need to drink a minimum of 2 L of water each day, in the form of clean water, or through the food we eat or in other types of liquid that contains water. This number is somewhat discussable—and is not necessarily scientifically documented.

All the countries mentioned in the statistics above have consumptions considerably above the numbers that UN regards as a minimum threshold for minimum water access—50 L per person per day. On national basis, strongly populated countries like China and India are well above the threshold, while Bangladesh and a number of countries south of Sahara in Africa are below the minimum numbers (Fig. 9.18).

New attitudes regarding consumption of water is a base for a sustainable development. The population in the world increases with about 80 million people every year. Improvement in social conditions also contributes strongly to the increased need for water. For example; The production of bio fuel has increased considerably in later years. Between 1 000 and 4 000 L of water is needed to produce one liter of bio fuel.

The consequence of the increase in production of bio fuel might be that the worlds need for fresh water increases with 64 billion m^3 every year [14].

A positive factor in the challenge to fight lack of water is that is good economy to provide clean water. WHO has estimated the economical advantages to halve the proportion of the people in the world that has not access to safe water within 2015, to be 8 times larger than the investments. The World Bank has made studies that 5 countries in Southeast Asia could increase their GDP (gross Domestic Product) by 2 % if they had better access to water and sanitary installations (Cambodia—7 %) [5]. The World Bank has estimated that Indonesia lost 6.3 Billion USD (2.3 % of GDP) in 2006, due to insufficient sanitary conditions and bad hygiene.

More than 80 % of the water that is consumed in the world is not cleaned or treated. WHO claims that cleaner water and sanitary conditions could have reduced the amount of diarrhea—illnesses by 90 %. They claim that the cause of death for 3.5 million people each year can be traced back to lack of clean water and unsatisfactory sanitary conditions in developing countries. In 2010 there were still 884 million people with unsatisfactory access the clean drinking water, and 2.6 billion people with unsatisfactory sanitary conditions. If we should have measured according to "western standards", the number of people without clean drinking water at home would probably be between 3 and 4 billion [5].

Of the more than 3000 cities in India, only 200 have partly or full sewage treatment systems. The upper limit for coli form bacteria in acceptable drinking water is normally100 organisms per 100 mL water. The Yamun River, running through New Delhi, receives approximately 200 mL of untreated sewage every day, and the coli form content increases to 24 million organisms per 100 milliliters. In addition comes considerable amounts of dangerous emissions from the industry [19] (Fig. 9.19).

Probably we too often define lack of water to be a problem for underdeveloped countries. The UN Climate Panel, IPCC, however claims that lack of water will be considerable challenge also in central and southern part of Europe in 2070. It is expected that summer ran will be

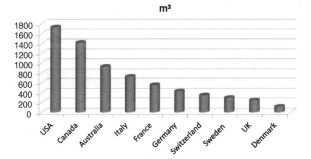

Fig. 9.17 Relative water consumption in various countries [17]

Fig. 9.18 Daily we regard water from the tap as an obvious commodity, but in some situations it might be useful to collect rain water to reduce the volume from the tap, or because the tap water is less accessible

Fig. 9.19 The clean water in the tap and the glass at our dentist is not obvious everywhere

reduced by 80 %, something that amongst others will have considerable negative effect on the hydroelectric production.

Already today, there are 120 million people in Europe that do not have access to safe drinking water. Even more people are lacking proper sanitary conditions [5].

A number of countries has far higher consumption of water, than what is generated in precipitation in own country each year A UN report [24] claims that 39 countries with a total population of 800 million people have half of their water consumption from sources outside their own borders.

As examples are mentioned Iraq and Syria that get most of their water from Euphrates and Tigris coming from Turkey. Bangladesh is dependant of that 91 % of the water consumption flows from India through the rivers Ganges, Brahmaputra and Meghna. Egypt is dependent on the Nile that comes from Sudan and Ethiopia. Water access and its quality are obviously also a potential conflict source.

Statistics from various parts of the world show a clear connection between the access to clean water and good sanitary conditions, and the life length. The tendencies are the same both for long time studies in well-developed as well as in less developed countries. There is also a good and clear development in the drive to provide the population in the world with good water conditions, but there is still a long way to go. A survey over the development in the world for the part of the population that lack good water conditions shows [20] (approximate numbers);

The world in total:	Reduction from 25 % to 17 %
Sub-Sahara Africa:	Reduction from 52 % to 44 %
East Asia and Pacific:	Reduction from 28 % to 22 %
South Asia:	Reduction from 28 % to 14 %
Latin America and Caribbean:	Reduction from 17 % to 8 %

We must accept that there are some dark numbers hidden in the statistics. One of the largest cities in India, Mumbai with 18 million inhabitants reports that 90 % of the

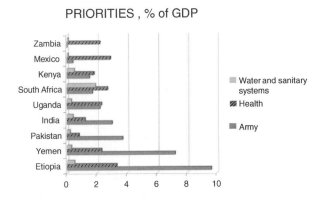

Fig. 9.20 Part of GDP (Gross Domestic Product) spent for various areas according to UN [20]

population has clean water. Half of the population, however, lives in so-called *zopadpatti*, along railway tracks, rivers and mangrove swamps. This part of the population is not included in the official statistics, and in these areas, the water and sanitary problem is considerable.

Priorities regarding water and sanitary installation vary from one country to another. A UN report from 2006 focuses on these priorities. The illustration below shows how various donor countries make priorities in how their funding is spent (Fig. 9.20):

How the supporting countries are making priorities for their support money to developing countries, varies strongly from one country to another. The same UN Report gives the following numbers for the part of the support aid that goes for water and sanitary installations. The numbers below only give some examples from the report:

Luxembourg:	9.5 %
Denmark:	7.8 %
Germany:	6.8 %
Japan:	6.2 %
Netherland:	4 %
Finland:	4 %
Sweden:	3.1 %
USA:	2.4 %
Norway:	2.4 %
Great Britain:	2.0 %

In addition, there are health and food consequences of the lack of enough water, and a number of other negative consequences. It has previously been pointed out how lack of water had led to problems and even total breakdown in fishing and agriculture. An obvious negative consequence and mass death of cattle, in Africa and South America are also found examples of how lack of water has led to less education, in particular for girls, because they are needed at home to carry water over long distances.

Even if an increasing part of the population in the world has got better access to water and better sanitary conditions, it is unfortunately so that the lack of water is faster than the increasing growth in the world population. An important reason for this is that increasing social standard increase the demand for water. Traditionally it was so that the agriculture was the largest water consumer, and in many cultures advanced irrigation systems has been utilized with rivers as the main water source. Possibly in particular, in the population rich countries in Asia there have hundreds of years been traditions for such advanced watering systems. With increasing wealth, increased water consumption goes to industrial and sanitary installations. Many places, this has led to increased contamination of water sources and changes in the potential supply in general. While the population growth increased fourfold in the last millennium, the water consumption increased sevenfold in the same period. This explains a great part of the water challenge in the word, and the challenge that has existed for quite some years now (Fig. 9.21).

The water challenges are definitely biggest in Africa south of the Sahara desert, Arabian countries and in Asia. The two countries in the world with largest population, China and India, that in a few decenniums will change place on the population statistics, has nationally enough water, but the water is very differently and unevenly spread throughout the countries. In for example China where 42 % of the population, 538 Millions, lives in the northern regions, while these regions only have access to 14 % of the national water resources [20]. The south western part of China, is the part of the country that has most precipitation.

An increasing part of the population is living in cities. This development will continue. The cities are nearly always without exemption located near a water source—sea, lake or river. This both ease and complicates the challenge with clean water and good sanitation systems. The concentration ease planning and water supply, but it increases the pollution potential (Fig. 9.22).

Though UN, the leaders of the world, set out 8 goals at the last chance of millennium—where drastic improvements should take place within the end of 2015:

1. Get rid of poverty and hunger.
2. Right for everyone for a basic education.
3. Stop sexual discrimination.
4. Reduce child deaths
5. Increase material health
6. Fight HIV and Aids
7. Secure a sustainable development.
8. Global development cooperation.

Fig. 9.21 The monsoon rain can be powerful in a number of countries that has lack of water. The lack of water is often due to lack of ability and resources to take care of the water that comes from the sky. Not seldom we see reports about both water problems from flooding, and water problems from lack of drinking water and lack of sewage systems from the same areas

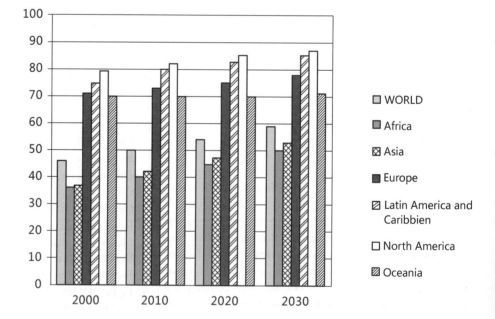

Fig. 9.22 The percentage of the population living in urban surroundings, according to UN [21]

Access to sufficient water and acceptable quality of all drinking water is a basic condition for most of these goals.

The former UN Secretary General Kofi A. Annan established the World Water Day in 2004 UN Council for water and sanitary questions. In the UN Human Development Report in 2006 [20], Kofi A. Annan amongst other says the following about the fundamental right for clean water;

> … The enormous numbers we use to discuss today's water and sanitation challenges must not be allowed to obscure the individual plight faced by ordinary people. This year's "Human Development Report" provides a powerful and timely reminder that the global water crisis has a human face: a child threatened with deadly bouts of diarrhea, a girl kept out of school to collect water or a mother denied opportunities to develop her potential by the demands of caring for relatives made sick by polluted water. The United Nations is deeply committed to this struggle. Access to safe water is a fundamental human need and a basic human right. And water and sanitation are at the heart of our quest to enable all of the world's people, not just a fortunate few, to live in dignity, prosperity and peace.

An economic factor that might contribute to decrease the access to clean water and sanitary wellness to the population in the world is the increased difference between the poor and the rich. The difference between the people that has a lot, and the ones that has less, increases in developing countries, and even in industrial countries. With increased water demand in many countries, people with high income will easily increase their consumption, with increased price of water as a consequence. This will decrease the possibilities for the poorest part of the population in the world.

According to UN [21], the control now is:

- The 1 % richest controls 50 % of the values
- The 5 % richest controls 71 % of the values.
- The 10 % richest controls 85 % of the value.
- The 50 % richest controls 99 % of the values.

UN appointed 2013 to be an international year for water cooperation, and the 22 March 2013 was appointed as the World *Water* Day. Mehta Kumar from India won the competition for the slogan for the occasion. This was presented by her at the UNESCO Headquarter in Paris 11 February 2013, and is: *Water, water everywhere, only if we share* [22].

Lack of water has of course led to disagreements around the world. The "fairy tale example" of how to handle such challenges in a rational manner is probably the water tribunal in Valencia in Spain. *Tribunal de Las Aguas* is historically a more than 1000 years old court system among equal citizens to solve disagreements between the farmers around Valencia about water rights in their fields. Today, this institution, that is one of the oldest democratic institutions in Europe, is more of tourist attractions than a reality. Once a week the tribunal meets at a central place in the city at Plaza de la Virgen, and takes decisions. Everything is oral, nothing is written on paper. A group of "equal men" sits in a ring and discusses, and takes decisions in water questions based on common sense, and everything goes on in openness to the public. On a fixed time very week they sit down on their seats. If there is nothing to discuss they raise and disappear. So tourists that come even five minutes late to watch the performance might be very disappointed.

An organization, calling themselves *Water Footprint Network* has made a manual for estimating the footprints from the water consumption in the world [23]. They remind that the consumer countries also must include the footprints all the way back in the production cycles. As example they mention that the footprints from cotton products from Malaysia also must include cotton products imported to Malaysia from China, India and Pakistan. In the calculation they operate with two types of water footprints;—"blue water footprints" regarding the use of clean water, and "grey water footprints" regarding contamination of water from the production cycles.

Water together with food and petroleum has been considered the top three strategic resources for sustainable development of the world.

References

1. Thaulow Sven: *Bad og Svømmeanlegg*, Norges Svømmeforbund og Norsk Cementforening, Oslo, Norway, 1965.
2. Internet 27.11.12: *Roman aqueduct*, Wikipedia, http://en.wikipedia.org/wiki/Roman_aqueduct.
3. Foss Bjørn: *Ferjelandet (The ferry country)*, Forlaget Nordvest, Ålesund, Norway, 1986.
4. Internett 28.11.09: The History of Concrete and the Nabataeans, http://nabataeaea.net/cement.html.
5. Internet 14.03.13: *Total World Energy Consumption 2010*, http://upload.wikimedia.org.wikipedia/commons/6/67/Total_World_Energy_Consumption.
6. The United Nations World Water Development Report, Executive Summary.
7. Flannery Tim: *Værmakerne (The Weather Makers)*, H. Ascheoug & Co, Oslo,Norway, 2006.
8. Wikipedia: *Desalination.* http://.wikipedia.org/wiki/Desalination.
9. Internet 23.02.12: *Water* http://en.wikipedia.org/wiki/water.
10. Noeleen Heyzer: *The task of providing drinking water.* China Daily, Beijing, Kina, May 27, 2013.
11. Kulkarni V.R.: *Concrete Sustainability: Current Status in India and Crucial Issues for the Future,* Concrete Sustainablity through Innovative materials and techniques, Roving National Seminars, Bangaloru, Jaipur, Nagpur, Kolkata, January 10–14, 2011, Indian Concrete Institute.
12. Malhotra V.M.: *Global warming and role of supplementary cementing materials and superplasticizers in reducing greenhouse gas emissions from the manufacturing of concrete.* Tenth ACI

International Conference on Recent Advances in Concrete Technology and Sustainability Issues, Seville, Spain, October 2009. Supplementary papers pp 421–440.
13. Internet 16.09.12. Water consumption Statistics, http://www.worldometers.info/water/.
14. Chapter 7, Residnets Living Level, Unified Statistics Paphlete for BRICS Nations, 2013.
15. Internet 16.09.12. www.drammen.kommune.no.
16. Statistisk Sentralbyrå (Statistical Central Bureau): *Statistiske analyser (Statistical analysis)—Naturresurser og miljø (Natural resourses and environment) 2005*, Statistisk Sentralbyrå, Oslo-Kongsvinger, Norway, December 2005.
17. Internet 16.09.12: SDWF- Safe drinking water—www.safewater.org.
18. Internet 16.09.12: Drinking water: http://en.wikipedia.org/wiki/Drinking_water.
19. Internet 12.10.12: Wikipedia: *Water wheel*, http://en.wikipedia.org/wiki/water_wheel.
20. UNDP: *Human Development Report 2006, Beyond scarity: Power, poverty and global water crisis,* United Nations Development Programme, New York, 2006.
21. World Watch Institute: *State of the World 2012.* Island press, Washington DC, USA 2012.
22. Internet 06.03.13: *2013—United Nations International Year of Water Corporation:* http://www.unwater.org/water-cooperation-2013/home/en/.
23. Internet 06.03.13: Hoekstra Arjen Y., Chapagain Ashak K., Aldaya Maite M., Mekonnen Mesfin M.: *The Water Footprint Assessment Manual,* http://www.waterfootprint.org/?page=files/WaterFootprintAssessmentManual.
24. Mary Bellis, Water wheel, Part I: http://inventors.about.com/library/inventors/blwaterwheel.htm.

Recreation 10

The closeness of water impresses us and is used for calming our mind, idea source, battery loading and experience through a many folds of alternatives and possibilities thorough our whole life and all over the world (Fig. 10.1).

Still water, a fishing pond, running brook, clucking water, sprawling water, frothing water, wave gush and ripple, the many-fold characteristics of water are great. Both ingenious designers and nature give us oasis of versatility and possibilities for recreational experiences.

The frontiers are not set by the water, but first and foremost by our own imagination, and to a certain extent by the architects' creativity to utilize, show and perform the cunning effect of the water on us. Direct, immediate impressions mix with associations and experiences both through our eyes, ears, and by direct contact, with water as temperature mitigator and it's finely divided cooling mist.

The quietness by the water and its noise absorbing effect can be as deafening as the roaring waterfall, the splashing from a great fountain and the sound of the waves on a beach. The silent noise from a small brook, a small well, or the clicking of the water in a gutter from a rainfall can transfer music in competition with the best made by human beings, whether you are digging one or the other mood of music (Figs. 10.2, 10.3, 10.4, 10.5, 10.6, 10.7, 10.8, 10.9 and 10.10).

Water also brings plenty of fun and pleasure to human beings, in particular children, all the time, in every season, and in various forms (Figs. 10.11, 10.12, 10.13, 10.14 and 10.15).

One of the structures on the in heritage list UNESCO is the Summer Palace, half an hour's drive from the centre of Beijing, China. Three quarter of the nearly 3 km^2 large area is covered by the Kunming lake built on basis of a blueprint of the West lake in Hangzhou. About half of the lake is natural, and the other half has been excavated. The water depth is hardly more than a meter.

The Summer Palace was built by Emperor Qianlong between 1750 and 1764 (Qing Dynasty 1616–1911). The Palace has been destroyed and rebuilt several times since its original building. The name "The Summer Palace" was given after the restoration of Emperor Guangxu from 1886 to 1895. Today the palace area is a very popular outing area, walking and jogging ground for the inhabitants in Beijing, and of course for tourists. In particular on Sundays, the area might be crowded. Even if the area has many attractions, many rush for "the long corridor" near the lake shore with more than 10 000 paintings on the beams and in the ceiling, very often with water motives.

The lake is located where the mountains starts north of Beijing centre, and where the flatland starts. The water in the lake comes from wells in the hills. Already during the Ming Dynasty (1344–1644), when the capital of China was moved from Nanjing to Beijing, the lake was utilized as a water reservoir for the capital, and many noble families built houses by the lake. A walk along the lake gives many fabulous impressions and experiences. One example is the arch bridge with 17 portals, which has been a model for park bridges many places in the world (Figs. 10.16, 10.17, 10.18 and 10. 19).

The Fountains

An important element in architectural water detailing is the fountain. The many fold of the fountain design is a kind of manifest of the attraction and the possibilities in water and

Fig. 10.1 Few experiences from the nature gives more calmness to our mind than a sunset by the sea, and possibly preferably near a tropical ocean

their attraction to human beings. Artists from all over the world have utilized the numerous sound possibilities and the music in water, often in a combination with attractive sculpturing, and sometimes accompanied by light and music.

Born at Majorstuen in Oslo, Norway, my first meeting and impressions from a fountain were the *Bear Fountain* at Majorstuen. With the curiousness of a child and the attraction of water, a strong hand from my parents was often needed to prevent another splashing in the fountain basin.

The fountain was made by the Norwegian artist Asbjørg Borgfelt (1900–1976), and was placed in its position in 1926.

The Majorstua crossing when the fountain was built is two different worlds of today. People nowadays are hasting past and are more focused in the traffic lights and the cars in all directions than the bear on top of the fountain and the small goat heads that are spitting out water. Traffic lights, cars and electricity liners are polluting the visual picture. When I as a small boy was passing the fountain, there were still horses in the traffic, and the cars had a different shape and speed. It was tempting for small boys to splash with the hands in the fountain, looking for fish:—it happened that someone put some inside, and to ask for permission to climb up to drink from the goat spring. Mother's reaction differed a

Fig. 10.2 From Norbulingka, literally the Jeweled Park, the successive Dalai Lamas' summer palace since 1780 in west suburb of Lhasa, Tibet, China, a popular meditation place and UNESCO World Heritage Site

Fig. 10.3 A pure early morning without a single of wind, only the sound of wave down the hill and early birds' song can be heard, composing a tranquil and vivid meditation music. Picture from a hotel by the seashore, Rushan, Shandong Province, China. July 2009

bit, but there many boys at my age got joyful contact with water in the Bear Fountain (Fig. 10.20).

My next fountain meeting was in the Vigeland Park. The main fountain was also used as a bathing place for children in the area in the summer before the first Norwegian outdoor pool came close by in 1951 (Fig. 10.21).

There are many famous fountains in the world, but personally Fontana di Trevi in Rome will always be number one. The personal thrill of the fountain started at Saga Cinema in Oslo a later winter night in 1956–1957 when some college friends watched the Swedish movie *Sjunde Himlen* (the seventh heaven) with the artists Hasse Eckman and Sickan Carlsson as the main roles. On their bus trip through Europe they amongst others visited Venice and later Rome. On the tram home from the cinema, the first plans came up on a hitch hiking trip to Rome in the summer of 1957. We had, by the way, seen the fountain previously in the American romantic comedy movie *Three Coins in a Fountain*—with the fascinating title melody by Victor Young with the same name, telling about the three coins in the fountain which bring happiness. The same fountain was later really made immortal by the Swedish actor Anita Ekberg as Sylvia in La Dolce Vita from 1960 bathing in the fountain.

So it became in the summer of 1957, 4 17 years old boys from Oslo, Norway went on a 5 weeks trip through Europe, with Fontana de Trevi as the final destination, two on motor bikes and the other two hitch hiking. Even if through life many fountains have been seen in many parts of the world, there is no doubt that this trip has been saved as the strongest and most lasting memories.

Fig. 10.4 A pure and crystal water arouses fanciful thoughts. Lotus Pond Park, Beijing, China. August 2009

The Trevi Fountain was completed in 1762 after 30 years of construction work and restored in 1998, where the rocks were cleaned and cracks repaired, and at the same time the fountain was modernized with recirculation pumps. However, the history of the fountain goes back much further (Fig. 10.22).

The fountain marks the end of one of the aqueducts from the Roman era that supplied Rome with Water. The aqueduct Aqua Virgo was leading water from the baths of Agrippa (Agrippa's Terme). The aqueduct structure was 22 km long and took water from a clean source 13 km away, and constructed about 29 BC. This was the sixth of the 11 aqueducts built by the Romans. When Gothic tribes conquered Rome in the middle of the 500's, they destroyed the aqueducts and the citizens had to take water from the infected water in the Tiber River instead. The water lines were later been restored in the 1400's, and became routine to construct fountains at the end of the aqueducts. The restoration of the aqueduct to the Trevi fountain was completed in 1453, but with a considerably simpler fountain than the one we see today.

In 1629, Pope Urban VIII found out that the existing fountain was not dramatic enough, and he asked Giano Lorenzo Bernini to make drawings for a new fountain. Even if the Bernini-fountain never was built, there are still many features from his original ideas. Before the present baroque fountain finally was completed in 1762, a number of Popes and architects were in the picture. The main architect seems to have been Nicola Salvi, but he died in 1751 before the structure was finished.

Rome can offer a number of fabulous fountain experiences, and some people have the opinion that the fountains at Piazza Navona are just as important and attractive as the Trevi Fountain. The oblong Navona-plaza was once a horse riding stadium with room for 30 000 spectators. Now it is a

Fig. 10.5 Harmonious coexistence of calmness and enthusiasm, Lotus Pond Park, Beijing, China, August 2009

Fig. 10.6 Japan's old capital, Kyoto in October. Water is a natural element in the 1000 years old gardens

Fig. 10.7 More Kyoto-early spring. The calming effect of water transforms good feelings

Fig. 10.8 Ken Domon-museum, Sakata, Japan. Water and swans and ducks are topics in many countries

tourist magnet packed with tourists and sellers of paintings and prints in all sizes and colors. The fountains in the south and north end if the plaza has its audience, but most attentions are probably given the Bernini-fountain about the four rivers—The Nile, Ganges, Donube, and Rio de la Plata in the middle of Piazza Navona (Figs. 10.23 and 10.24).

Fig. 10.9 There are some Asiatic moods in the water art at the Trollstig plateau, at the mountain road in Romsdal, Norway. The National Norwegian Road authorities have in later years designed and built a number of resting areas along the most scenic Norwegian roads, with design that everyone might enjoy. The Trollstig plateau is one of these, and the water art design is to be happy from

Fig. 10.10 Hot water drips down to the mini-pond on top of the ice mound, splashes under the sunshine and forms a crystal ice mound under freezing winter temperature. January 2009, in a hot spring resort, Chicheng, Hebei Province, China

Fig. 10.11 Summer fun in a Sunday morning in a modern shopping centre in Beijing, China

Rome is a "wonder-city" for fountain walks. It is not far between the water holes whether you prefer one or the other meaning of the word (Fig. 10.25).

However, these days you do not have to travel as far as Rome to find urban splashes and water holes (Figs. 10.26, 10.27, 10.28 and 10.29).

Fig. 10.12 An intimate contact with friendly water creates a lot of fun and experience when climbing up along the Gudong Forest Falls, Guilin, Guangxi Zhuang Autonomous Region, China, August 2009. *Photo* Fuyin Chen

Again and again we are fascinated by the ability of good architects to utilize and appease pictures through the simple medium—water. The ability to emphasize the basics or position of a structure is decisive for the total impression.

There might be a lot of water sculptures in a bridge too. The centre of Prague was put on UNESCO's World heritage list in 1992. A natural part of this is the Charles bridge over Moldau (Vitrya). Prague was one part of the Austrian province Böhmen, and the capital for the Böhmian kingdom, and the old part of the city was established in the 1200s (Fig. 10.30).

But it is not only the architects that create fascinating details. A walk on the rocks near the sea can show many details created by the yearlong wave action if only you use your fantasy. The Grandmother and Grandfather rocks (Hin Ta and Hin Ya on Koh Samui in Thailand) are indeed very special. The legend tells that an older couple drifted in and formed the rocks on the beach (Figs. 10.31, 10.32, 10.33, 10.34 and 10.35).

River Cruise

Over the last 10–20 years, recreation in the shape of river cruises has gained considerable popularity. Today there is hardly any large river that does not offer a well organized vacation possibility. The catalogue from a random Norwegian travel bureau that has been specializing on the topic gives not less than 28 river alternatives with possible visits to 32 different countries

Fig. 10.13 Stone skipping, a recreation of boys in the pond under the natural water fall, Mt. Juyu, Rushan, Shandong Province, China. July 2009

- Amazonas, Brazil
- Bhagirathi (Ganges delta): India
- Chobe: Namibia, Botswana
- Dnepr: Ukraine
- Danube: Romania, Bulgaria, Serbia, Croatia, Hungary, Slovakia, Austria, Germany
- Dordogne: France
- Douro: Portugal, Spania
- Elben: Germany, Tjekkia
- Ganges: India
- Garonne: France
- Gironde: France
- Guadalquivir: Spain
- Guadiana: Portugal, Spain
- Gøta kanal: Sweden
- Irrawaddy: Myanmar
- Mekong: Laos, Thailand, Cambodia, Vietnam
- Mississippi, USA
- Mosel: Germany
- Nilen: Egypt
- Oder: Pollen, Germany
- Po: Italy
- Rhinen: Netherland, Germany, France
- Rhone: France
- Seine: France
- Saone: France
- Svir (+Neva): Russia
- Volga: Russia
- Yangtze: China

On several of the rivers cruises are offered with various alternatives with respect to length and luxury, and sometimes combinations of several rivers are offered. Other catalogs offer river cruises on rivers also in North and South America and on other continents (Figs. 10.36 and 10.37).

The 83 km long trip on the Li river, or Lijiang, where the river is surrounded by colorful mountain sides, green hills, small rice paddies, sleepy grassing or bathing water buffaloes, duck farms an fishermen on bamboo floats—is both a fascinating experience and recreation, and is like sailing through a colorful painting. An old Chinese poem says about the landscape that; *The river looks like a green silk garland, and the hills are like emerald hairpins.* Lijiang later runs into the Pearl River, the third longest in China, and the largest river in Southern China (Figs. 10.38, 10.39, 10.40 and 10.41).

Boat life is for many people the first thing they think about talking about water and recreation.

Norway has comparable to other countries a very long coast line, probably leading to the fact that people in Norway are more focused on boats than many other places.

Boat life is a versatile subject in itself and in the mind of the typical boat people. It spans from a trip in the canoe to slow rowing for fishing in a small boat, motoring slowly along small beautiful islands in the sunset, water skiing in the fiord, sailing on a board or a small raft, or vacation in the sailing boat with anchoring up in a small sound. The menu is many fold and very detailed for the many boat lovers (Figs. 10.42, 10.43, 10.44, 10.45, 10.46, 10.47 and 10.48).

Boat life might also spend a few days, or even weeks on board a cruise ship. We have previously mentioned river cruises, but a cruise trip to exotic oceans attracts many as an

Fig. 10.14 A "Battle" of water, Water-Splashing Festival of Dai Nationality to celebrate the bumper grain harvest, prosperous people and livestock. Now tourists are welcome to join the celebration, among which children are the happiest and most active. Xishuangbanna, Yunnan Province, China. July 30, 2008. *Photo* Hao Sui

interesting vacation alternative. However, also many find the Norwegian coast and fiords as attractive experience possibility (Figs. 10.49 and 10.50).

In Norway, most people have another international recreation alternative—*Norwegians goes hiking in the weekends.* There are many hiking alternatives. Some people love a slow walk along the streets or the local road. Some spend the Sunday to climb a mountain top, and some like a trip barefoot an a small beach. For particularly many, number one hiking alternative is a trip in the local forestation either on foot in the summer months or on skies in the winter. The experience from small lakes, brooks, rivers and moors is an important part of this recreational hiking experience (Figs. 10.51, 10.52 and 10.53).

Some hikers combine their trip with fishing. **Fishing** is not only a recreation alternative, but pure obsession for some people as well. The many fold is considerable—from a trip with a simple rod and some worms to a little lake, to heavy and advanced equipment for fighting the salmon in a river where the costs often exceed the price of salmon in the shop and where the catching itself makes other recreation alternatives and impulses secondary.

Fishing is also to walk on the cliffs along the ocean with a rod, sitting in a boat in an early morning or late afternoon

Fig. 10.15 A Lake at suburb of Beijing, by a lakeside in a park after a late snow in early spring, very rare and thus arouses interest and fun of visitors who made this. Marble carving combined with snow ball showcasing a better combination. March, 2012

with the line over your pointing finger. Fishing is kids laying flat on a pier after eagerly looking for fish or crabs. Fishing is slow motion of a motor boat with line in the raft in a quiet Sunday morning. Fishing is also with your small fishing boat in the morning to look for the results in the nets overnight. Fishing is passion for some but indifferent for others. But, fishing is also one of the most important industries in long coast countries like Norway (Figs. 10.54, 10.55 and 10.56).

A trip to ocean side is an important recreation alternative, not only in the summer (Fig. 10.57).

Town people hiking alternative can be a trip down to harbor or the river to watch the passing ships or to experience other life along the water way. In the largest harbor city of the world, Shanghai with a population of 23 million. The Huangpu River is an attractive alternative. A combination of the walking street of Nanjing Road with all the shopping centers open on a Sunday leading to The People Park with the river by the Bund and views to the finance centre and all the new futuristic tower buildings in adds much to the attraction (Figs. 10.58, 10.59 and 10.60).

Spa

About 20–30 years ago, Spa was something connected to special cities in Europe, and to a certain degree to other continents when mineral rich hot or cold springs could be found. The Spa-sites offered health cures of variable types. The offer was normally looked upon as relatively exclusive, costly and for rich people (Figs. 10.61 and 10.62).

About 10–20 years ago, hotels and resorts in many places of the world began to present offers about Spa. Today we find Spa offers in all variations spreading as an epidemic all over the world—in small and large hotels and resorts, in basements, in shopping centers and many other thinkable and non-thinkable locations. Spa has become an offer for everyone, and not only for the more wealthy part of the population as before.

Fig. 10.16 A panorama picture from the Kunming lake. It is windy and the waves are considerable in the shallow lake. The arch bridge with the 17 portals on the other side of the lake, to the *left*

Fig. 10.17 A close view of 17-portal arch bridge, Summer Palace, Beijing, China. March, 2009

Fig. 10.18 Water details are important architectural features. From Tjuvholmen, Oslo, Norway. Running water gives creative details in a new and busy city life

Fig. 10.19 Water details—hotel in Thailand

Fig. 10.20 The Bear Fountain at the Majorstua crossing in October 2012. It seems like the bear is wandering why the water is not there—"why do I have to stand here and freeze when there is no water anyway"

Fig. 10.21 The main fountain in the Vigeland Park. Without water a grey October day, some of the attractions disappear from the structure

Fig. 10.22 Fontana de Trevi. A legend tells that if you throw a coin in the fountain, your return to Rome is secured. We visited the fountain and took this picture 56 years after our first visit. The biggest difference was that on the first visit we were nearly alone, but on the last visit, it was packed with visitors around the fountain

Fig. 10.23 Piazza Navona, seen towards north with the Maure Fountain by Giacomo della Porta from the 1570's in the foreground

Bath from health-bringing springs and bath for enjoyment and recreation are very old phenomena, but the word **Spa** originates from the Belgian town Spa in the Ardennes hills. The town had its great era in the 1600's when people from many places in the world came there to cure various illnesses. Utilization of these wells or springs, however, started much earlier. The casino in Spa was established in 1763, and is said to be the oldest casino in the world. (Could it be that the hundreds of thousands of spas we see around the world also might end up with a casino offer?)

Someone has claimed that Spa is a shortening of the Latin words *Sanitas Per Aquam* (health through water), while others have claimed that is historically incorrect.

Many places in the world have histories with bath from special wells or springs,—hot or cold, and through several thousand of years, these springs have been claimed to be not only health bringing, but also part of special religious cleaning rituals. Such cleaning rituals might found in most religions. As mentioned in Chap. 9. *Water consumption,* the Romans, 2000 years ago, used baths for social activity,

Fig. 10.24 Details of the centre fountain of Piazza Navona

Fig. 10.25 A 2000 year old water hole at the height above Foro Romano, Rome, Italy

recreation and pleasure, and the Romans learned it from the Greek. The city of Bath in England was one of the most Northern of the many baths that the Romans founded. The water there is also mineral rich and healthy by earth heat. The Roman name of the city was *Aquae Sulis*.

The procedures in the original Spa's vary considerably. They drink mineral rich water, and what are often included, —hot and cold baths, massage of various kinds of wrapping in blankets or mud or clay, are often combined to a local specialty.

One of the oldest and most exclusive bathing resorts we know is the hot spring called *Huaqing Chi* on the north slopes of the Lishan mountain in Lintong, about 25–30 km from Xi'an in China. The resort building was started by King You during the Western Zhou Dynasty (about 1100–711 BC). Later the castle and the resort were extended by the first Chinese Emperor Qin (259–210 BC) when the capital of China was established in Xi'an. The resort was further extended by Emperor Wu during the Western Han Dynasty (206 BC–24) [1] (Fig. 10.63).

Fig. 10.26 Water mirror and zigrat in Asker community. Asker is a smaller coastal community, some 20 km west of Oslo, Norway. The community is situated along the Oslo Fiord, but the community centre is situated some kilometers from the sea. The cafés along the artificial water are popular meeting places in the summer

Fig. 10.27 In the middle of October the water is emptied from the fountain for the winter. The interest for the plaza is reduced

Fig. 10.28 The fountain and the ziggurat in winter dress. The interest for the white water as sliding object is considerable from the small children

Fig. 10.29 An attractive fountain can also be simple. Some water flowers and splashing water do the trick (Thailand)

Fig. 10.30 The Charles bridge over Moldau in Prague. The building of the bridge started in 1357, and it is built by Bøhmian sandstone. The bridge as we see it today comes from the end of the 1699s/beginning of the 1700s. During the big flood in Moldau and Prague in August 2002 they were afraid that the bridge might be destroyed by all the debris floating against the pillars, but the bridge survived the forces. In a concrete technology sense, the bridge is of special interest, because the history tells that the farmers around Prague had to supply lots of eggs during the construction for mixing into the mortar between the sands to be blocks

 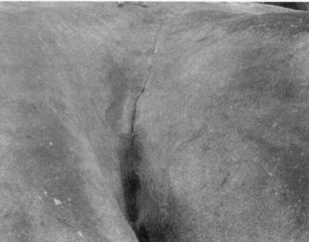

Fig. 10.31 Hin Ta and Hin Ya (grandfather and grandmother)-rocks, Lamai, Koh Samui, Thailand

Fig. 10.32 The edge of the hotel pool is visually floating in one with the sea below. The wharf on the other side of the bay adds to the realism in the picture. Vivanta Hotel, Willington Island, Cochin, Kerala, India

Fig. 10.33 From Kira Maya Resort, with the nature reservation Hat Yai in the background,—an inland hotel in a mountain area about 2 h drive north east of Bangkok, Thailand. Use of water is an important tool in the landscape architecture

Fig. 10.34 The ability of water to mirror light contributes to making the Nahu lake a popular goal for a night walk for the inhabitants of the old Chinese capital Xi'an

Fig. 10.35 There are a lot of recreations in an overnight trip in an old rice barge too. This boat has 8 double tourist cabins, and offers an intriguing experience on the Chao Phraya-river, Bangkok, Thailand. The calmed, quiet life on the barge gives a fascinating contrast to the busy, pulsating life on and along the river

Fig. 10.36 There are plenty of variations, many-fold possibilities and fascinating experiences and time for quiet contemplation in a riverboat trip on the Danube river. The 2850 km long river runs through 10 countries and can offer numerous variations in trips, varying in length, ship size, and quality

The name of today, Huaqing Chi, originates from Emperor Xuanzong extension in year 723 during the Tang Dynasty (618–907), who spent large sums of money to make the resort into a luxury palace, also called the Summer Palace of the Emperor. The resort is found a number of artificial pools from hot springs. The hot Huaqing baths with about 3000 years of history include some fascinating episodes in Chinese history from ancient times and up to today. Most well known is probably the love stories between Emperor Xuanzong and his concubine Yang Guifei who was said to be one of the four most beautiful women in ancient China. The Emperor himself should in his 41 years of reigning have visited the resort 36 times.

Emperor Dowager Cixi from the last Chinese Dynasty-Qing (1644–1911) was hiding in the Huaqing baths when Beijing was occupied 1910, and the resort was also the

Fig. 10.37 A boat trip down the Li River from Guilin in south China (culture centre in Guangxi since the Northern Song-Dynasty for over 1000 years ago) to Yangshuo, is one of the most important tourist attractions in China. Guilin is each year visited by 20 million tourists in total and 2 million tourists from abroad

Fig. 10.38 The kitchen in back of the tourist boats on the Li River prepares food for tourists, and there is no doubt that it is busy. The food is good even if the look of the kitchen gets variable characteristics from the tourists

Fig. 10.40 A night view of the river in Guilin

resident of General Chiang Kai-shek in 1936 (Figs. 10.64 and 10.65).

Swimming

We do not know for sure when human beings started swimming. It has been registered that Sumer's 5000 years ago built a pool of 60 × 30 m, but we do not know for sure that someone have been swimming in the pool.

Roman poets tell that 2000 years ago swimming was a popular sport among the Germanic tribes in the north.

It is claimed that King Karl the Great (742–814) was a master swimmer, and we have in Chap. 7 mentioned that Egill Skallagrimson in Iceland 1100 years ago had to swim 10 km to survive.

Thaulow [2] tells that already in 1965 there were more than half a million swimming pools in USA. No wonder that USA has been a massive medal collector in swimming in the Olympic Games and in world championships.

Only after the Second World War the activity of building swimming pools and indoor pools in particular got important effect in Norway. It became a national goal that the whole population should be able to swim. With a comparatively long coast many people had close nearness to sea and water,

Fig. 10.39 It is close between the tourist boats on the Li River, a vacation day

Swimming

Fig. 10.41 River cruise boats ready for next week's trip on the Duero River in, Gaia, Portugal

Fig. 10.42 Small boat harbor in Strømstad, Sweden, in September—the summer and the season is over. The Norwegian boats that occupies and fills up the harbor in the season, have left for more homely waters. The inhabitants in Strømstad "likes" Norwegians, because the their visits are the backbone in the economy of the city, but they are also relieved that the summer is over

Fig. 10.43 Strømstad. On the other side of the harbor, there are still plenty of boats. Most of them will be there over the winter

Fig. 10.44 In October, many put their boats on land for the winter, and the square meter value of the land in those storage areas is high over the winter

Fig. 10.45 But there is a lot of fun in small boats too!

Fig. 10.46 An adventure which you can control the rubber raft only a little bit with the stick in hand when drifting in Fir Wood River, Shibing, Guizhou Province, China. August 8, 2012. *Photo* Hao Sui

Fig. 10.47 Life and recreation potentials in the various archipelagoes have an important place in the heart of Scandinavian people. Many have their own boat, or they use the public boat transport. Strømstad, Sweden; The ferries are ready to take the local population and tourists to the popular Koster Islands

Fig. 10.48 A trimaran trip in the sunset at Mauii on Hawaii can also have a high rating on many people's wishing list. The picture is from the finish of San Francisco-Hawaii Race in 1976, with the finish line at Laihaina on Mauii, Hawaii's old capital

Fig. 10.49 The city of Stavanger, Norway, has got a visit from the famous "Queen Elisabeth", one of the most legendary passenger ships in ocean history, now a cruise ship

Fig. 10.50 The bottom of the Geiranger Fiord in late June. The fiord might be "crowded" with cruise ships in the summer months

Fig. 10.51 The parking lot at Lke Sem, Asker, Norway, a Saturday in October. The Asker Community offers many hiking alternatives. A trip around the Lake Sem is one of the most popular ones, and it might be difficult to find a parking spot in particular in the autumn and spring

Fig. 10.52 In many places the community has made it easier for the hikers where wet areas have to be crossed. The picture is from an Asker moor in October

Fig. 10.53 But, there might be some wet challenges in the wet autumn season

Fig. 10.54 Bjørn, happy on the forest cabin stairs after the catch of the evening

Fig. 10.55 Lofoten Island, Norway in December. No recreation but hard work in catching the cod

and the survival argument was very important. Swimming as recreation and safety measure is important to most people, but swimming has never got the image of a national sport in Norway in the same way as for example, skiing, or like in countries such as Australia.

Many find that the ultimate swimming experience is to swim in the ocean, but swimming in an indoor or outdoor pool might be quite pleasant and recreational too.

The Norwegian Swimming Association tells in the "anleggsplan" (Construction Plan) [3] that up to January 1, 2010 there were 204 short-lane pools (mostly 25 m) and 5 long-lane pools (50 m) in Norway that have been built with subsidies by gambling money from the government, but none of these pools is up to international competition standards. Many of them are starting to get old, and need considerable renovation. According to the swimming

Fig. 10.56 Ice-fishing has its own charm, for large as well as for small. The Stein Fiord, Norway

Fig. 10.57 But in October there is not much beach life in Nordic countries. The ducks have got their territory back. There is not much boat life either. Only the ferry from Oslo to Denmark is cruising out the Oslo Fiord

association, there is a need for twice as many swimming pools than what can be found today.

The Norwegian Swimming Association claims that a reasonable public goal is to have at least one 25 m swimming pools for every 10 000 inhabitants (Figs. 10.66, 10.67 and 10.68).

Scuba Diving

For many, diving might be regarded as a rather special form of recreation. However, it has become a popular hobby, recreation, and adventure experience for many thousands worldwide. When the Earth is more than ¾ covered with

Fig. 10.58 It is crowded along Huangpu River on a Sunday in November, Shanghai, China

Fig. 10.59 Photography along Huangpu River on a November Sunday, with thousands of other Sunday hikers around

water and has similar comparative variation in topography under the water as above, and the life in water might be as versatile as that above water, there are many that have missed something of the wonders of our planet when they have tried to swim with a compressed air tank to investigate the sea bottom more closely. Most of the fun and life can be found on rather shallow depths. Most species of fish, shellfish, and corals are found on depths of less than 20–25 m. As you go deeper, the light is reduced, and with it more and more of the spectacular colors can be found. The aphoristic zone is area in the oceans and lakes that are so deep there is too little light for the photosynthesis of the plants. This zone is deep closer to equator due to the light refraction. Diving in tropical waters is far more spectacular than for example in northern waters like in Norway. However, this does not mean that you will lack fantastic diving experiences in colder waters. Tropical dives compared to dives far from equator differ mostly on three points: the color difference is nearly like the difference between color and black and white television, the number of species of all life and corals might be many fold in the tropical waters, and the colder waters far from equator require much heavier diving equipment (Figs. 10.69 and 10.70).

Fig. 10.60 The view from the Bund to Pudong in Shanghai is fascinating. Pudong has some of highest skyscrapers in the world and also the television tower, the Pearl of the Orient. Shanghai Tower is under construction when the picture was taken. When finished, it will be the second highest building in the world with 632 m high

Fig. 10.61 Smiling hostess presents Spa-offer at Raddison New World Hotel, Shanghai. (*left*)

Fig. 10.62 Lauras Spa in a basement, Asker, Norway. Here mature Thai women give Thai massage at a price comparable to 10 times of that for a similar massage in Thailand (*right*)

It cannot be hidden that a well planned trip below the ocean surface needs both equipment and technique (Figs. 10.71 and 10.72).

Golf

Any respectable golf course has water as part of the landscape architecture and as obstacles in the golf play (Figs. 10.73, 10.74, 10.75 and 10.76).

The bathing beach has its definite attraction, and possibly this is the water recreation activity that gets most votes when the summer comes or during a trip to more tropical waters in the cold part of the year (Figs. 10.77, 10.78, 10.79, 10.80, 10.81, 10.82, 10.83 and 10.84).

To visit a river or probably even more a waterfall might be recreational by itself. To hear the boulder, feel the water

Fig. 10.63 Water from above and water from the ground. The May rain is flowing down over the Emperor's Summer Palace, while it also bubbles up from the ground

Fig. 10.64 The marble statue of Yang Guifei is a popular photo object

Fig. 10.65 In small bathrooms the visitors in Huaqing Chi can enjoy water with a temperature of 43 °C. The mineral rich water is claimed to have health-bringing effect, and a therapeutic effect on the skin through the mineral and organic compounds. The bathrooms also have massage benches

mist and to see the fantastic formation and the forces, or to experience the play of color in and above the water and in the structures around will fascinate most people no matter the waterfall is small or big (Figs. 10.85, 10.86, 10.87, 10.88 and 10.89).

The ultimate recreation is by many called the vacation. Many have their vacation dreams connected to some kind of water. Water in the shape of rain or lack of rain is one of the important criteria evaluated when the vacation period will be evaluated afterwards. Rain in good or bad way can change the vacation picture completely (Figs. 10.90 and 10.91).

Water has, through centuries, also been used for "anti-recreation" purposes. Walls have been the main tools used to keep something such as people, thoughts, earth and stone out or inside. The world's most famous wall, the Ming Wall is officially 8851.1 km long, where 6259. 6 km

Fig. 10.66 "The Water Cube" (178 m^2 and 31 m high), National Swimming Center for Beijing 2008 summer Olympic Games, designed based on the spread pattern of cells and the natural structure of soap bubbles, also a model for water saving that all backwash water is filtered and returned to the swimming pools, saving another 140,000 tons in recycled water a year. December 2008

Fig. 10.67 Landøya swimming hall in Asker, Norway, a beautiful, very popular and well visited public pool built in the 1970's. It consists of the main pool of 25 × 12.5 m and a therapy/children's pool of 12 × 8 m. The community of Asker with a population of about 50 800, expected to have 62 500 inhabitants in 2020, has two swimming halls

Fig. 10.68 Typical corrosion attacks in a swimming hall. Rust staining on steel doors and from reinforcement in the concrete. Water is recreation, but it also sets its footprints in the breakdown of the structures. From Landøya swimming hall in Asker

Fig. 10.69 A glimpse from corals and fish life in the South China Sea and Adaman Sea

Fig. 10.69 (continued)

Fig. 10.70 The first meeting with a shark is always an adrenalinitic experience, but most types of sharks are rather harmless. The picture is a from harmless Reef Shark, Gili Gili islands, Lombok, Indonesia

Fig. 10.71 A Norwegian television hostess makes the equipment ready together with a marine biologists. The divers are ready for making television shots for a program about the Tjuvholmen artificial reef project in Oslo harbor (mentioned at the end of Chap. 11). The project has aimed to clean the water and bring sea bottom life back in the harbor

Fig. 10.72 Ratcha Yai, Thailand. It is 28 °C in the water, and a lighter dressing is possible in warmer water. Many of us claim that this increases the recreation potential compared to diving in colder waters

Fig. 10.73 A first long drive over water on the first hole is a challenge for recreation players. This is from Kjekstad Golf Club, Røyken, Norway. It is a clear Wednesday morning in October. The frost on the ground from the night is still present when the weekly senior citizens tournament starts. The season goes towards its end. The first challenge of the day is to get ball over the small lake in front of the tree

Fig. 10.74 Small landscape pearls might be found on the golf course. Hole number 7 on Kjekstad Golf Club, Røyken, Norway is a par 3 hole where you have to drive across a small lake with this colorful island

Fig. 10.75 This lake has to be crossed on hole 10, Kjekstad Golfklubb, Røyken. A nice autumn day for golf in the middle of September

Fig. 10.76 Beautiful water details on hole 5, Phuket Country Club, Thailand. Light rain and 35 °C temperature do not reduce the recreation pleasure

Fig. 10.77 The ocean and white beaches attract millions of people for battery loading every year. Kata Noi beach, Phuket, Thailand. Water temperature is about 30 °C

Fig. 10.78 Vacation in Thailand. A trip with a "longtailboat" along the beaches is a definite recreational attraction

Fig. 10.79 Not only people enjoy playing in the water. From a swimming pool in East Troy, Michigan, USA

Fig. 10.80 And not much water is needed for the water buffaloes the enjoy themselves, Tamil Nadu, India

Fig. 10.81 It looks like dolphin is smiling during its play in the water

Fig. 10.82 Both small and large aquariums are recreational attractions at home, restaurants and in various museum options. This is from a Japanese restaurant in Beijing China

Fig. 10.83 When the dolphins play and show their act, they seem to have fun, but this also attracts an audience

Fig. 10.84 It is in early spring in Norway. There is not so much open water yet, but the nice weather melts the snow on the railway station roof. The water splashes down, and a duck couple is attracted by the chance for fresh water

Fig. 10.85 Not all waterfalls are using the power for electricity production. In Iceland they have decided the famous Gullfoss (the Gold fall) shall be kept as an experience of nature for recreation and tourism instead of hydroelectric power production

Fig. 10.86 Gullfoss, Iceland-seen towards the mountains in the north east

Fig. 10.87 Jade-like water pouring down, Niagara Fall, Canada, April 2004

Fig. 10.88 The water falls over the Trollstigen serpertein road in Møre og Romsdal, Norway give a strong contribution to the tourist attraction

Fig. 10.89 Along the Geiranger fiord in Norway, many waterfalls drop directly from the mountain ridge more than 1000 m directly into the salt water fiord

Fig. 10.90 Rose Garden in rain, Bangkok, Thailand

Fig. 10.91 Rose Garden some hours later after the rain, Bangkok, Thailand

Fig. 10.92 From Hradcany Castle in Praha, Tjekkia

are walls, and 359.7 km are ditches which often filled with water. The rest is natural obstructions [3]. An interesting story about the construction of the Ming Wall at Jiayuguan near Beijing goes like this: The work needed thousands of stone blocks with a dimension of $2 \times 0.5 \times 0.3$ m which had to be transferred to the top of the mountain. Workers, enlightened by heaven, built a road and sprinkled water in winter and made an efficient icy way to transfer the building materials up to the top.

Water channels or ditches around castles and prisons are other examples (Figs. 10.92, 10.93 and 10.94).

Fig. 10.93 Cherry blossom recreation walk along the water channel circling the Emperors Palace, Tokyo, Japan

Fig. 10.94 We end the chapter by a sunset over the ocean. Kata Noi, Phuket, Thailand RECREATION!

References

1. Internet 28.05.13: *Huaqing Hot Springs.* http://www.travelchinaguide.com7attractions/shaanxi/xian/huaqing.htm.
2. Thaulow Sven: *Bad og Svømmeanlegg*, Norges Svømmeforbund og Norsk Cementforening, Oslo, Norway, 1965.
3. Norges Svømmeforbund: *Anleggsplan for svømmeanlegg – Mai 2010,* Internet 21.03.13, www.svomming.no.

The Food Source 11

Oceans, rivers, and lakes are probably much smaller food sources than they could have been if we had been a bit more intelligent. Only about 1 % of our food comes from water if we count this in energy values. If we count this in protein values the figure is a bit higher. However, researchers are of the opinion that the figures could have been 10–20 times higher if we consider evaluation based on the photosynthesis as the bases for all food production, and consider the area in the ocean and take into account the reduction in production efficiency in water due to light refraction.

If we only look at the protein available for human food in the world, fish, or rather sea food, makes out about 10 % of the consumption in the world [1]. In total, the food production from the water makes out a bit more than 100 million tons, somewhat increasing in line with more efficient seafood farming. Of course this does not include food produced on land with the use of agricultural water, which takes much higher share for the food of human beings and raised animals (Fig. 11.1).

The reduction in sea food from maritime fishing after the turn of this century can be recognized in sharp contrast to the strong growth in the previous 50 years (Fig. 11.2):

- In 1950 the total production was less than 20 million tons.
- In 1970 the production passed 40 million tons.
- In 1990 the production passed 70 million tons.

In the years 1996–2002 the catching of all the 4 fish categories with more than 10 million tons of annual catching stagnated. The catching of cod and herring was reduced with 20 % in the same period [2].

The catches in the world oceans have since the turn of this century been on the way down due to overfishing and reduction in the resources for more than half of the types of fish having the largest quantity. In line with the increased efficiency in catching methods and tools, the volume has been reduced simultaneously. Some claim that we are destroying the ecological balance in several oceans. The consumption of sea food is stable or slightly increasing due to a considerable increase in sea food farming. The largest increase in the sea food farming we find takes place in Asia and in particular in special types of shell fish.

While Japan in many years was the world's largest consumer and producer of sea food, the picture has changed over the last 10 years or so.

China is today the world's largest consumer, producer, importer, and exporter of sea food. This was not the situation in the 1990's (Figs. 11.3 and 11.4).

A number of Asian countries have in later years got high up on the list of sea food producing countries. In 2010, 57 countries were recorded by FAO with a production of more than 200 000 tons. These numbers do not include fish farming (Fig. 11.5).

The North Atlantic Ocean has been one of the richest areas for fishing in the world, and accounts for about 20 % of the world's marine fishing. Even if the number of fish types is less than in typical warmer waters, the number of types is considerably high, and higher than many might expect. In the Norwegian book *Våre Saltvannsfisker (Our Salt water fishes)* [3], is mentioned 161 species of salt water fish and 173 species including shellfish in the North Atlantic. According to the book, fish has over time got a more limited meaning than in older times. Then everything in the sea was fish even if it was worms, starfish, seals or whales.

The World Watch Institute report for 2012 [4] says with reference to FAO (UN's Food and Agriculture Organization) that sea food farming in 2009 reached 38 % of the world's consumption and that 11 million people were employed by this industry. If we are going to maintain the part of the world diet that comes from sea food, we have to produce 40–60 million tons more than we produce today, and most of this must come from sea food farming (Fig. 11.6).

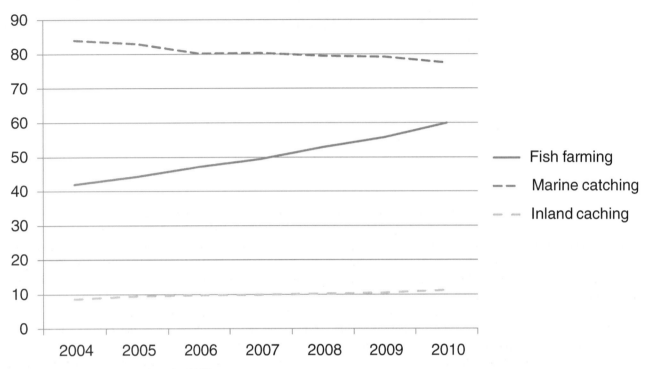

Fig. 11.1 World production of sea food [33]

Fig. 11.2 A tempting summer meal with cold sea delicatessens is always something to look forward to

Over the last 70–80 years the number of fulltime fishermen has been drastically reduced also in Norway. In 1930, there were 130 000 fishermen with membership in fisherman working organization. This number was reduced to around 70 000 in 1960, and in 1980 the number drops below 30 000. From the turn of this century the number has been reduced to well under 20 000 [5]. Today there are just above 12 000 active fulltime fishermen.

The number of persons working in fish farming in Norway today is close to 5000 [6].

According to the newspaper Aftenposten with reference to ICES (International Ocean Research Council) [7], the amount of the various species caught in the Arctic Sea is (Fig. 11.7):

- Macrel: 2.9 million tons.
- Hering: 7.9 million tons.
- Blue Whiting: 2.3 million tons.

Export of fish and fishery products from the 3400 km long Norwegian coast has long traditions, and there are quite a few successful stories of special products from this industry.

Omega—3 fatty acids, which is today a fashion expression in the diet and health food debate. However, the smell of the cod liver oil liquid might be a problem for many (Fig. 11.8).

Peter Møller was born in the town of Røros on 26th of April, 1793, but already at the age of 16 he was employed as pupil by the Chemist Gottweld in the town of Kristiansund.

Fig. 11.3 At the end of the summer season, the water level is low and you have to go far into the river for fishing. From the Sichuan Province in China. The large catching volumes in China spans include large area in catching methods and fish farming both in the sea and on land

Fig. 11.4 Development in total production—marine catches plus fish farming. According to FAO 2003. China, Indonesia and India have strong increase, Norway has a small increase. Other nations have decreased their production

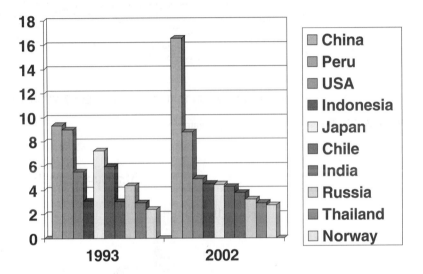

In 1819, he travelled to the Norwegian capital Kristiania (now Oslo), and was employed by the Chemist store Svaneapoteket. He got his exam as pharmacist in 1822, and in 1829 he bought Svaneapotekeket and together with other company Lilleborg Klædesfabrkk. This factory which he took over as the only owner in 1842 later becomes the main base for his development for steam melting of cod liver oil as medicine. In 1853, he made his method known along the coast, and he built cod liver oil cookeries with new equipment at 3 locations [8].

The first year they produced 20 barrels of cod liver oil. When Peter Møller died in 1869, there were more than 70 cod liver oil factories along the Norwegian coast that used his method.

Møller's tran (cod liver oil medicine) became one of Norway's first brand name products, and even if the green bottles is looked upon with some fright by many, later yellow capsules and other omega-3 products have become a natural part of the breakfast for many not only in Norway, but in a great number of countries around the world as well (Figs. 11.9 and 11.10).

Presumably, it was the Spanish fishermen in about the year 1 500 that invented the salted and dried fish (in Norwegian *Klippsisk, from the cliffs they were dried on*) as a preservation method when they fished by new found land. In 1690's the Dutchman Jappe Ippes that settled down where Kristiansund is located today, and started the salted and dried fish production with the royal prerogative. Speed in the production, however, first came in 1737 when the Scotsman John Ramsay bought the fishing village of Grip. Later came more Scots into the game, which resulted in that Kristiansund became the Norwegian Clipfish capital for over

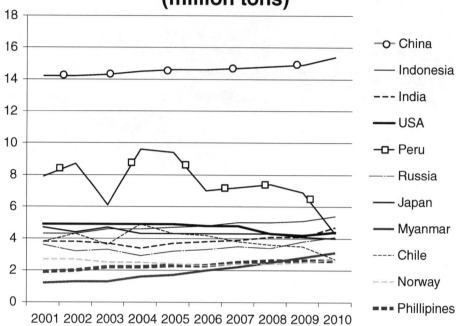

Fig. 11.5 The largest sea food producing countries in the world [33]

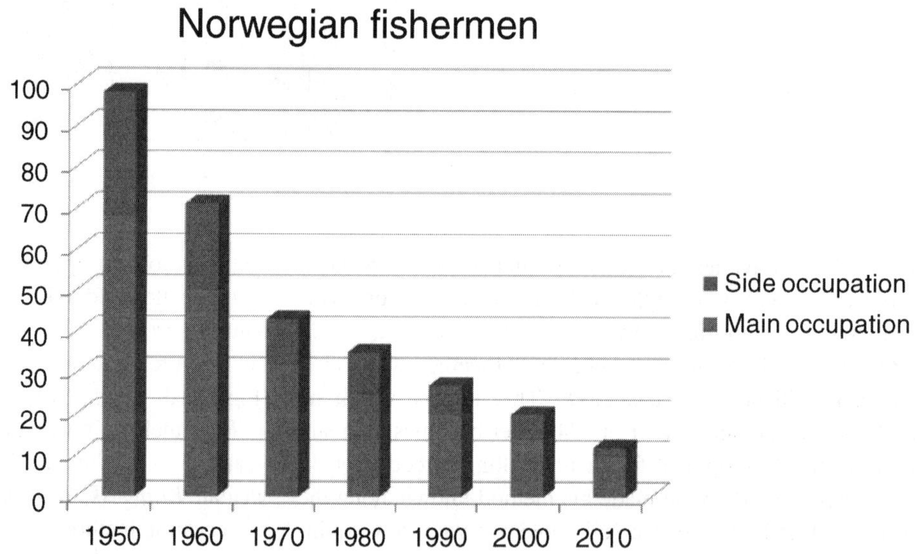

Fig. 11.6 Norwegian fishermen (1000) [6]

200 years. After the World War II the town of Ålesund and other places along the coast become more dominant in the salted and dried fish exports [9].

Next to Iceland and the Faroe Islands, Norway is among the largest salted and dried fish exporters in the world with Portugal, Spain, Italy, Latin America, and Mexico as the major consumer countries.

While the **stockfish** is unsalted and dried out there in the wind. The fish that should be made to dried and salted cod is cut along the back bone, and then boned out, salted and placed in stacks under salt maturation. Originally the drying happened on the cliffs, but today it is more common with the indoor ripening and drying. Dried and salted cod (Clif-fish) is preferably produced from cod but

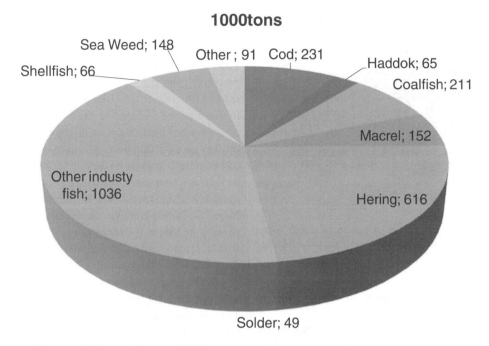

Fig. 11.7 Norwegian fishery catch 2004 according to SSB [5]

Fig. 11.8 The range of omega-3 products might be considerable in a Norwegian Chemist store

Fig. 11.9 The green cod liver oil bottles have been placed on the lowest shelf in a normal food store in Norway now, and you might need a guide to find them

haddock and other fish types have also been utilized (Fig. 11.11).

Whaling has been an important part of Norwegian development history. Whaling has been happening all along the Norwegian coast. However, it is perhaps the catch in the Southern Ocean and the whaling County Vestfold, which is most strongly remembered in people's minds in Norway. Today, the whaling is a debated topic and in many countries both banned and viewed as a dirty habit and a great tragedy. In the 1950–1960's, however, to be whaler was much like an honorary designation to recognize. Little did we think over the fact that Norway was one of the foremost nations of a predatory catching that led to the result that most whale species have been considered to be endangered. We had little consideration for natural balance, and was mostly concerned with increased capture efficiency and increased prosperity and social development, and west of the Oslo fjord is an example in particular. The contribution to a formidable over-exploitation was hardly mentioned and much less understood (Fig. 11.12)..

The whaling history is old. Stone pictures from 6000 years B.C. from South Korea indicate that Stone Age people were hunting whales with boats and spears. Written sources tell about bay whales catching along the Norwegian coast in the 800's. Catching with harpoon was normal in the 1200's. Whale catching on the Faeroe Island was at least practiced from the 900's. And whaling in Iceland can be dated 1100 years back. Use of harpoons in whale catching by hand started in Japan in the 1100's [10].

Even when the modern whale catching started with the use of the grenade harpoon, it was mainly based on costal catching from stations on land. The ship owner Svend Foyn had already been a pioneer with respect to seal catching, and started whaling from 1864 in the northernmost county in Norway—Finmark. Foyn combined three factors: the harpoon, a grenade in the harpoon and a new type of whale ship with a steam engine. The first steam engine whale ship was *Spes and Fides* that was built for the cost of Svend Foyns at a wharf Christiania in Norway in 1864 [11]. In 1863, he designed the first grenade harpoon with an explosive head. In 1873, he got patent on his invention and a monopoly on the whale hunting in the Finmark County until 1882. From the beginning of the 1880's a small fleet of boats from Vestfold County was whale catching in Finmark based on licenses from Foyn. The license conditions would probably today be regarded as outrageous, but Foyn earned a fortune. The profit from the massive hunting was sale of whale oil, which in England was sold for 20 lb per ton. The catching was, however, far from sustainable, and from January 1904 the new Norwegian whale hunting law forbids whale hunting outside the counties of Nordland, Troms and Finmark [12].

The pelagic catching of whales with expeditions with a large floating cooking ship and up to 10 whale catching ships had started earlier but really took off in the 1950's. The hunting mainly took place in the South Sea around Antarctic. In addition to Norway the whale hunting was also done by ships from Japan, England, and Netherland.

It was the whale oil that was the bases for the whale industry. Only a few countries were interested in the whale meat. However, some countries like Japan have long food traditions with whale meat. The lack of food during the

Fig. 11.10 In the Chemist store the variation and offer of omega-3 product is considerable

Fig. 11.11 Fish is salted and sun-dried, a small fishing village in Mt. Dayu, Hong Kong, China, where people have a long such tradition. August, 2009

Fig. 11.12 A Sandefjord—souvenir anno 1951. A gift on my father's 60 years birthday from a proud schoolmate who was a whale hunter from Sandefjord. The whale even smile. Sandefjord was once the whale capital in Norway

Second World War made whale meat an important protein source. If whale meat is a delicatessen or not might be debated among those of us that had to eat it as youngsters. Some mothers, however, were able to reduce the fish taste of the steak by introducing a lot of onions (Fig. 11.13).

International cooperation about the regulation of whale hunting started in 1931. The International Whale hunting Commission, IWC, was established in 1946, but it was not until 1982 when IWC issued a moratorium with abandon of all commercial whale hunting from the season 1985–1986. It really came to the result after a number of meager hunting years that we got a stop in the hunting.

The international debate about whale hunting or not today is characterized by both rational and emotional arguments from both sides. The fact, however, is that most of the various whale species are threatened by destruction, while for some there should be possible to have a sustainable hunting. One of these types are bay whale, where the source in the North Atlantic is estimated to 110 000 whales [12]. Norway has set the limit for hunting to 1000 bay whales per year, but annually only half of this is hunted. In the 1950s the bay whale catch was close to 4000 whales. A research

Fig. 11.13 In the center of the town of Sandefjord we find a number of things that remind about the whale adventure in the middle of the last century

article from 2012 [13] claims that the economical output of this catch only is about 18 million NOK/RMB.

Canned fish has been an important Norwegian industry and export item, and has set its footprints along the coast from Sponvika Canning at Idde Fiord at the border to Sweden in the south-east and all along the coast to the west and north to the border to Russia. From the famous Christian Bjelland's sardines in olive oil to fish balls, cod eggs, and caviar, the invention activity has been great until the frozen fish adventure entered the market and took over the importance of the canning industry (Fig. 11.14).

Norwegian export of fish is about 30 billion NOK/RMB per year. The research institute SINTEF has estimated that in 20–30 years this can be 8–10 doubling.

During all history there will always be someone that want to jump higher, run faster or do things more efficiently than others, and so it should be. It is a part of our nature. But, in sport and in most areas in the society, we make rules against cheating and so that improvements will not come into conflict with healthy competition and a sustainable development.

To improve the working condition for fishermen is in itself a sustainable action. The original idea of trawling for fish is part of such improvement, and hardly anything to blame anyone for. But the trawling in all its varieties is still protected, which is strange and not responsible and sustainable to the opinion of many. The nightmare picture of someone picking flowers in the field with a tractor might be comparable with several incidents in the trawler industry. Pictures in the media show effect of bottom trawling where the trawl nets have destroyed coral colonies and made deep trenches in the sea bottom. Trawling of endangered species is another non-sustainable action in the industry. A third negative example is the dumping of the fish from the trawl net that does not have the same market value as other species, and that might reduce the allowable quota that has been awarded.

A UN-report [14] claims the trawling fleet of the world fishes up to 2.5 times the volume that gives a sustainable future for the fishing industry, and that so-called industrial fishing should be terminated to the advantage of the coastal fishing fleet.

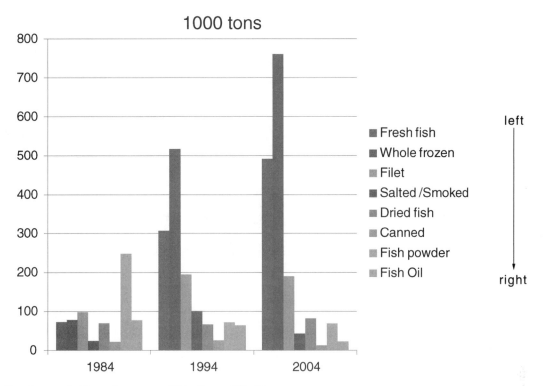

Fig. 11.14 Development in Norwegian export of fish. *Source* SSB [5]

Researchers have clearly documented that there is enough nutrition for the fish in the seas and oceans and other types of sea food, and that the oceans can contribute even more to the food supply in the world, but the catching must take place in a sustainable manner. The world statistics clearly shows that some types of catching methods, trawling in particular, are not sustainable activities. Cowardice politician, being too afraid of the economic machine behind the activity, must be the only reason why such fishing methods have not got a termination plan.

A Norwegian book by Lena Amalie Hamnes—*Naken uten fisk (naked without fish)* [15] with good scientific references also claims that trawling uses 3–4 times as much energy, and emits 3–4 times as much CO_2 per kilogram of fish as the coastal fleet.

Fish Farming:

We have over the last 20 years or so seen an impressive increase in the fish farming industry. The Norwegian fish farming industry includes a number of sea food species, but it is the salmon and trout farming industry that is dominant. It is the largest in world of its kind (Fig. 11.15).

The research institute, SINTEF has estimated a possible onward growth potential for the Norwegian fish farm industry of 4 % per year to a five doubling from 2010 to 2050. However, there are quite a few conditions that have to be fulfilled to reach such growth. The most important obstacle is to find enough feeding for the fish. The plant based part of the food cannot be increased much, as this will decrease the fish quality. The most exciting option is to harvest plankton from the oceans. The resources are present, but new technology is needed [16].

For a number of countries, freshwater fishing in rivers and lakes is considerable, and as a worldwide total, fresh water fishing makes out nearly 10 % of the total volume. In countries like Bangladesh, Burma, Uganda, Egypt, and Chad, sea food from freshwater sources is a very important part of the total food source [17].

The river fishing in countries like Norway is relatively modest, but it is an important food source and income for the once involved.

In addition comes the unrecorded hobby, fishing, which is estimated to about 10 000 tons (Fig. 11.16).

The catching methods for fish around the world are many and vary considerably from one part of the world to the other. Some methods are rather "exotic" to most of us.

Cormorant-fishing is one of these "exotic" methods, but it hardly represents the heaviest catching volumes. The method, however, is probably one of the most sustainable. This method has been utilized in China and Japan for more than 1300 years, and it is still in use (Figs. 11.17, 11.18 and 11.19).

The fishing takes place when the fisher normally from a small bamboo raft is taking advantage of a Cormorant bird.

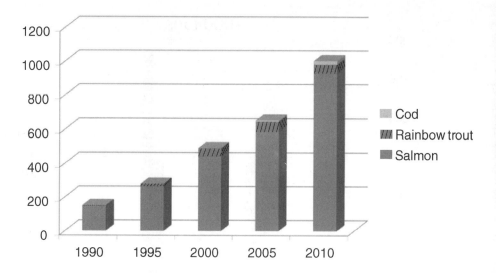

Fig. 11.15 Sale of Norwegian fish farm fish in 1000 tons [6]

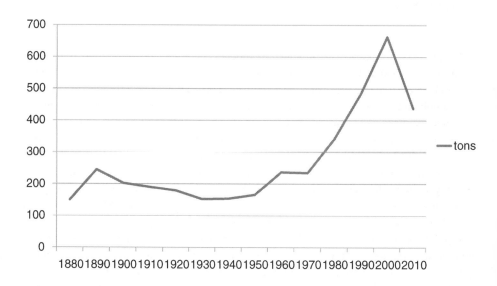

Fig. 11.16 The development in Norwegian river fishing, salmon and trout (tons) [6]

The bird is a good diver, and has a very long neck. At the lower part of the neck the fisher has attached a string. The bird dives for fish. When the fish is caught, the fisher tightens the string. The bird jumps up on the raft and spits out the fish.

The catching method is used both at day time and nighttime, but it seems like catching with a lantern after dark is most popular. The method also has old tradition in Macedonia, and has been tried in several European countries.

According to Wikipedia [18], there is documentation showing cormorant-fishing in Peru in the 5th century and long before the use in Asia, so the idea might be originating from different parts of the world.

The richness in species varies considerably in the various oceans. The number of species is higher in warmer water and in particular in areas that have not experienced any ice ages. In total there are recorded over 20 000 species of fish in the world. The highest number we find in the oceans around the countries is in the South China Sea and the Andaman Sea. This is true both for fish and corals (Fig. 11.20).

The great richness in species is also reflected in the range of **prices** for different types of sea food. **Abalone** is on the very top of the price list for sea food, and price above 500

Fig. 11.17 Cormorant fishing perpetuated by the Japanese artist Kesai Eisen (1790–1848) in a woodblock print in the series 69 stations on the Kisokaido over 150 years ago. Eisen was one of the great Japanese ukioe artists. The Kisokaido-series was started by Eisen, but completed by the master of the time, Ando Hiroshige

Fig. 11.18 Cormorant fishing at the Li River near Guilin in China

NOK/RMB per kg is not rare. The total world production is below 2 million tons per year. Abalone is a type of snail with shell only on one side, where the snail attaches itself to flat rock or stone or plate on the other side. Most of the production comes from farming with a typical sales size of 6–8 cm. There seem to be 17 different types of abalone around the world, mainly in tropical and tempered waters. For example in Europe, the European type lives in waters south of the channel between England and France. Depending on the type of abalone, the size at the time of sales varies from 6 to 25 cm in width of the shell and up to 7.5 cm in thickness. The abalone is also sometimes called an ear shell due to the shape of the shell. The large muscle foot of the abalone is recognized as extremely good food in a number of countries. Commercial fishing of some size might be found in countries for example California in USA, Japan, Mexico, China, and South Africa. The largest type of abalone is found in USA, and might be 30 cm long. Normally the abalone lives in the coastal zone at shallow depth. It fastens to rocks and boulders and moves at night to find food like sea weed and algae.

A growing problem is that the abalone has enemies, such as star fish and crabs. In some coastal areas some poisonous algae might also be a problem. Japanese researchers claim that the optimum depth for farming is at 4–10 m depth. This is under the algae belt and not so deep that lack of light which reduces the growth. Both due to the limited area in the ideal depth and to avoiding the enemies of the abalone snail, some years ago they started farming in Japan on floating sinkable platforms. After that farming factories on land have produced abalone babies to a size of 1–3 cm in size, they are set out on the platforms. The abalones are placed in net boxes with plastic plates to grow on, and they are fed with seaweed. Also the food is produced through farming (Figs. 11.21 and 11.22).

The farming platforms are fed and harvested in a surface position, where after water is pumped into the corner tanks and the platform is sunk to 4–7 m depth, an ideal depth for growth under the algae belt and with acceptable light condition (Fig. 11.23).

The Norwegian newspaper VG on 6th of January, 2013 [19] reports about a new price record for the sale of blue finned **tuna** at the fish market in Tokyo. The restaurant owner Kiyoshi Kimura paid 10 million NOK/RMB for a fish of 222 kg. This was 3 times as high as the previous record. The article referring to Associated Press claims that 80 % of what is caught at this type of tuna is consumed as sushi in Japan. It further claims that this type of tuna is threatened by

Fig. 11.19 Cormorant fishing can also be a fun for tourist, Elephant Trunk hill looks like an elephant that is drinking water, Guilin, China. August 2009

Fig. 11.20 A good fish restaurant in the old part of Shanghai offers fresh fish in many variations, many of them unknown for Europeans

Fig. 11.21 Abalone farming from a floating sinkable platform, Ohfunato, Japan 2000

Fig. 11.22 The growth of abalone, farming at the east coast of Honshu, Japan (år = year)

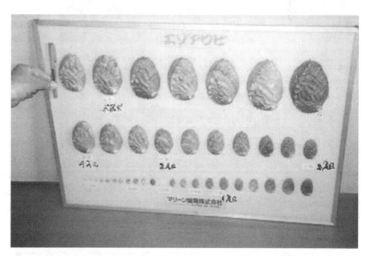

overfishing and that the source has been considerably reduced both in the Pacific ocean and the Atlantic Ocean, and that WWF (World Wildlife Fund) is asking people not to buy blue finned tuna (Fig. 11.24).

The tuna fish is the fastest swimmer in the sea with velocity up to 75 km/h for the yellow finned tuna. Tuna is found in a number of variations in the tempered zones in all the oceans of the world. In general it is anticipated that tuna lives from 45° north to 45° south. Tuna is part of the mackerel family, where the smallest species can grow up to 50 cm in maximum size. The largest is the Atlantic blue finned tuna that has been measured up to more than 4.5 m and a weight of over 600 kg. It is believed that the Atlantic blue finned tuna might be up to 50 years old. Average length of the various types of blue finned tuna is 2 m.

Traditionally tuna was caught by line and rod, and the catches for the five most popular species grew from 1940 to the middle of the 1960's from 300 000 tons to 1 million tons. Over time also net catching has come into use and the total catches are now up in 4 million tons per year. Of this 68 % is in the Pacific, 22 % in the Indian Ocean and 10 % in the Atlantic Ocean [20]. The large increase in the catches has led

Fig. 11.23 Abalone boxes are fed with seaweed from local farming

Fig. 11.24 Blue finned tuna is a delicatessen both for the eye and on the sashimi plate. The tuna pieces melt when they arrive on your tongue. The picture is from a restaurant in Shanghai

to a situation that some species are threatened by overfishing. In particular this is the case for the southern blue finned tuna.

Farming of tuna fish has been initiated several places in the world, for example, Australia, Vietnam, Sri Lanka, The Maldives, the Mediterranean Ocean, USA and Japan (Fig. 11.25).

Frogs, and in particular frog legs are on the menu in a number of countries, while in others it is regarded too strange to eat. Frog farming is also going on in quite a few countries, and frogs are an important export article with Indonesia and China as the largest exporters and France, USA, and Belgium as the largest importers. International trade with frog legs is said to be about 40 million USD per year, and it is claimed that about 3.2 billion frogs are eaten in the world each year [21], or in average one frog per every second inhabitant in the world. In Europe, it is first of all France that has been recognized by its frog leg dishes. However, it is also tradition to eat frogs in countries like Spain, Portugal, Croatia, Slovenia, and Greece. The French food tradition has also spread to the southern part of USA and frog legs are an important New Orleans—delicatessen.

In Asia we find Frogs on the menu in a number of countries like China, Indonesia, India, and Thailand.

The nutritious frog legs taste like something in between fish and chicken. In the Liaoning province in Northeastern China they serve the whole frog caught from forest, glazed as a delicatessen and it is said to be good for your love life. When the small, black, about shining frogs are sitting looking at you from the serving plate, it is hard to find what to associate them with some kind of chocolate perhaps. In some religions like for the Jews and Muslims, frog is not allowed as food.

Sea sausages or sea cucumbers are not a part of the diet in many parts of the world, but in China and Japan it is

Fig. 11.25 These small creatures make some impressive sounds that could be heard at a distance of several 100 m

considered a nutritious and healthy sea food. This creature is found in more than 1000 various varieties around the world, with over 30 in the North Atlantic. Normally they become 10–30 cm long, but some species can be up to a meter long. In addition to being used as food it might also be used as medicine in a dried version.

Sea cucumbers farming are found among others along the coast of Northern China (Figs. 11.26 and 11.27).

Of all the different food sources from the sea, we must not forget the **sea plants**. Annually we harvest about 7 million tons of sea plants. Of this, about 1/3 goes for various industrial purposes and about 2/3 goes for human food, where countries like China and Japan are the largest consumers. Farming of sea plant is about 6 times as large as the harvesting of wild plants. Farming of sea plants, however, is rather labor intensive.

In countries like Japan, sea plant has been an important daily part of the diet. Around the world sea weed has over the last few tens of years become a natural ingredient used in the preparation of sushi (Fig. 11.28).

China has recorded over 100 factories that refine seaweeds. More than 80 % of this goes for food.

Norway has particular good conditions for harvesting of sea weed along the long coast, and is the largest harvester in Europe with about 3 % of the sea plant harvesting in the world, only second to China.

A large continuous belt of the sea weed called large brown algae (latin: Laminarea Hyperborea) is by many

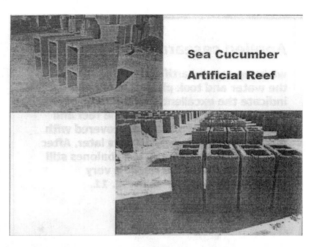

Fig. 11.27 Mass production of concrete components for the farming of sea cucumber in Liaoning Province in China

called a sea weed forest. These forests might be found from 1 to 30 m depth with its major in a depth of 10 m.

Use of sea weed in food and for animal food has been known in Norway for more than 100 years. Burning of sea weed was an important industry along the coast already in the 1700's and the 1800's. Most of the ash was exported to Great Britain for use in glass production and for production of soap. In the last years before the production was terminated in the 1930's, the sea weed ash was used for iodine production.

Fig. 11.26 Sea sausage or sea cucumber might be an important ingredient in a Chinese meal. It is served in many varieties. Here in a soup type dish from Beijing

Fig. 11.28 A sushi chef in work to make a portion

More than thousand persons were employed in the sea weed burning industry in Norway [22]. When it was discovered that there were iodine in the ash in the beginning of the 1800's, the export grew to about 6000 tons of ash, comparable to more than 100 000 tons of fresh sea weed. The acid smoke from the sea weed burning gave root to one of the first environment debates in Norway [17] (Fig. 11.29).

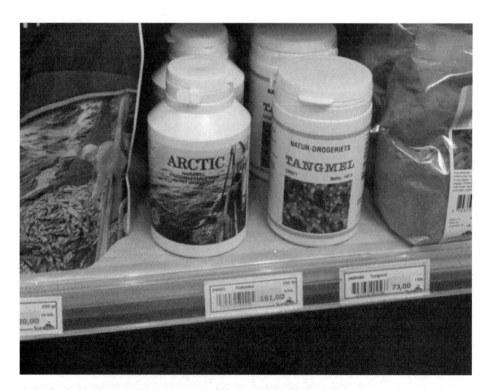

Fig. 11.29 Sea weed powder has its natural place in an assorted Norwegian health food shop

Many of us know sea weed powder as an interesting healthy addition to the diet, but it is used as addition in industrial production and is mostly used in many countries. The alginates from sea weed used as a thickener for color-paste in printing are the largest market segment. The alginates are also used as a thickener in various foods as marmalade, yogurt, mayonnaise, etc. [23].

The Norwegian alginate industry started based on hand-picked sea weed as a raw material source. Mechanized harvesting of the algae started in 1964. The fully mechanized trawl that is utilized with a capacity of about one ton per trawl was developed in the 1970's [22]. There are regulations in Norway for this activity to ensure a sustainable harvesting operation.

Another and perhaps a bit more exotic "algae—water—health—product" but with a similar look produced from fresh water algae in warmer areas in the world is "Spirulina." A modern expression for the product is *Arthrospira*. The product is mainly produced in Asia, Africa, and South America. Spirulina was used as a food product by the Azteks in middle America already 500 years ago [24].

According to Wikipedia [24], Spirulina has positive tests as health products in a number of places in the world.

With the enormous technical development we see in China today, it is probably not so strange that a very technical and health wise advanced production of Spirulina takes place in China. Close to the Chenghai Lake in Yongsheng in Yunnan Province, in southwestern China, they have found good conditions both for the health positive green algae for Spirulina, and the more expensive type containing astaxanthin. The production capacity of Chenghai Bao Er Co. is 220 tons per year, but ongoing investment of 1.7 billion RMB will increase the production considerably through a new line with a larger capacity of 1500 tons per year.

Mr. Tan Guoren, owner of the company when meeting with the authors also mentioned an ambitious plan of identifying the possibility to capture the CO_2 emitted from a new dry process cement plant with a capacity of 3000 tons of clinker per day located 15 km away, which can "provide" far more than enough CO_2 for the production of algae. The question again comes to the cost for capturing CO_2 which can be suitable for algae cultivation. The production of 1 ton of algae needs 2.3 tons of CO_2, as Mr. Tan introduced. Small as this mitigation might be in terms of the huge volume of CO_2 emission, it nevertheless can be a good demonstration of sustainable CO_2 capture, storage, and utilization (CCSU) (Figs. 11.30 and 11.31).

There are many of the food alternatives from the sea that most people never get a chance to experience. One of these alternatives is turtles and turtle soup, even if turtles have been a human food alternative, they nearly as long as people have lived on the planet. Researchers even claim that the oldest species of turtles are 215 million years old. The turtles are cold bloodied and most of them live most of their lives on land. Among the 263 various species of turtles we find both salt water turtles and freshwater turtles. Unfortunately, most of the species and on most continents are threatened by extermination. Consequently, many countries have restrictions for sale of food from turtles and for the shell of turtles. The many-folds in turtle species also give a wide variation in size. It is claimed that the largest types can have length up to 2 m and a weight of more than 900 kg. Prehistoric turtles

Fig. 11.30 Production facility for Spirulina in large basins near the Chengai Lake in Yongsheng in Yunnan Province, China, and examples of some of the products from Lijiang Chenghai Baoer. August, 2013

Fig. 11.31 Director Mr. Tan Guoren in Chenghai Bao Er Co. is proudly showing the plan for the extension of the Spirulina facilities. August, 2013

have been recorded with length up to 4.6 m. The smallest species are found in South Africa and are less than 8 cm in maximum length.

Turtle soup and turtle stew are still found on the menu and as a delicatessen several places in the world. The shell on the bottom side is also an important ingredient in traditional Chinese medicine.

Many places in the world we find farming of turtles both as a possible food source and in an effort to try to increase the amount of wild turtles. It is claimed that more than 1000 turtle farms are found in China (Fig. 11.32).

Another of the specialties from the sea—oysters, might be daily food in some parts of the world, but very special and only for the grand occasions. Oysters were farmed in England and France already during the Roman time some 2000 years ago, but somehow the interest in Europe nearly disappeared for nearly 1½ thousand years. Traditionally this was cheap and typical poor people's diet, but that is not the case anymore. Oysters have a very variable shape, size and taste around the world, as nearly 100 different types might be found. In some countries with warmer waters oysters are nearly recognized as troubled weed, because the shells attach to boats tools, quays, and structures and are more for trouble than for joy. Oysters from farming are normally ready for eating after 3–4 years, while wild oysters might be 20 years and up to 20 cm long (Fig. 11.33).

According to FAO [25], the production of European flat oysters *(ostrea edulis)* has been considerably reduced since the top year 1961 with 30 000 tons to under 4 000 tons in later years. The reason is partly attributed to that the market has been taken over by the round Pacific Ocean oysters *(crassostrea gigas)* being 3–5 times cheaper in the wholesale market. The delicious European flat oysters now only make out only 0.2 % of the world total market for oysters. Hence, this type has been reduced to an exclusive delicatessen in the high price class. In Europe, France and Spain represent more than 90 % of the production.

Amongst others along large areas of the Japanese coast you might find intensive oysters farming. One of the main centers for the activity is the small island of Miyajima

Fig. 11.32 Young turtles at a farm at the Grand Cayman Island in the Caribbean Ocean. (When Columbus found the Cayman Islands he called them Tortugas (turtles)). This turtle farm was started as a private farm in 1968, but was taken over by the government in 1983. Every year the farm sets out several thousand mature turtles into the sea. This species of turtles can as grownups have a weight of 100–300 kg. The most productive female at the farm has managed to lay 1799 eggs in one year. The females normally lay about 100 eggs each time. They dig a hollow in the sand, where they lay the eggs, and then cover them with sand afterwards. The eggs have many enemies. The females try to fool the enemies by digging many false hollows

Fig. 11.33 Norwegian wild oysters enjoyed in the village of Homborgsund on the south coast. *Photo* Elisabeth Jahren

outside Hiroshima. The island has its annual oysters festival in the second week of February each year (Fig. 11.34).

On Miyajima you might enjoy a number of variations in serving of oysters, but most popular is probably; oysters naturally and "Oysters Rockefeller." "Oysters Rockefeller" is served all over the world, but the recipes vary considerably. The origin comes from the restaurant *Antoine's* in New Orleans in USA, and it is claimed that the original recipe was a secret. However, it contains a sauce from different vegetables and herbs, spinach, and bread crumb, and it was then baked in the oven. Some places the "Oysters Rockefeller" will be gratinated with cheese.

The oceans hide a number of creatures that are relatively unknown on most tables. One of the creatures that in total have the largest total biomass is **krill**. Krill is a small ocean animal with 86 different species. It lives in all the oceans of the world, and has a look like a small shrimp. First and foremost krill has a function as a link in the food chain for a number of the animals and fishes in the sea. The amount of krill in the oceans is estimated to 500 million tons of biomass, or nearly twice what is the case for the human beings

Fig. 11.34 The well known 16.5 m high landmark Otorii Tori in Miyajima

in the world [26]. Krill is used both for human food and for medical purposes. In Japan Krill is called *okiami*. The annual catching of krill that mostly takes place in very southern waters amounts to 150 000–200 000 tons per year, which up to now has been regarded as a sustainable quantity. However, some researchers have now started to question this. Reports about the negative effect of the climatic change on the amount of krill both in southern and northern waters have appeared. Tim Flannery, with reference to Angus Atkinson in British Antarctic Survey [27], claims that 60–70 % of the amount of krill on the southern part of the world can be found in the Southern Ocean. Investigations in the period from 1926 to 1939 and from 1976 to 2003 tell that there is not recorded any increase or decrease of the amount in the first period. A completely different pattern is recorded after 1976. Since then the amount of krill has had a strong decrease of nearly 40 % per 10 years. It looks like the amount of krill changes with the amount of ocean ice in the Antarctic the previous winter. The amount of ocean ice has been drastically reduced after 1950, and the northern border of the ocean ice has moved from 59.3° south to 60.8° south with a reduction in the volume of 20 %. Flannery claims that the climatic changes are serious threats against the world's most productive ocean and against the animals that live and feed there.

Sometimes it might be reasonable to be reminded about that the sea is not only a food source for people (Fig. 11.35).

The incredible variations in species in the sea get further accentuated by the fantasy in how we process, store, and prepare the sea food alternatives. The sea food many folds and possibilities are impressive to everyone with an open mind (Figs. 11.36 and 11.37).

The fish markets as a food source or just as a tasting or visiting object are fascinating in particular in coastal towns. The often quick speech from the people behind the counter adds to the attraction of fish markets as tourist visiting objects (Figs. 11.38, 11.39, 11.40 and 11.41).

The presentation of the sea food is uttermost important. Japanese food traditions say that 70 % of our taste comes from the eye, 20 % from the nose, and 10 % from the tongue. The first step is obviously therefore to tempt the customers, and few places in a food center are more tempting and exiting than the sea food department (Figs. 11.42, 11.43, 11.44, 11.45 and 11.46).

Talking about water-food, the natural thought is seafood. However, as mentioned in the chapter about water resources, water is a necessity also for producing food on land, and globally 70 % of the water consumption goes for agriculture. The world's most important and by far the largest food consumption item is rice. In most places of the world, rice production is not depending on artificial watering. Flat rice fields come under water in the rainy period when the rice is planted, while the rice has it growth period and is harvested

Fig. 11.35 The ducks havve occupied the fishing lake. There is no space for the fishermen with their rods. Sakkata, Japan

Fig. 11.36 Drying of fish is a well know preservation method in many countries. Fish and octopus on the roof of the barge is not abnormal in Aberdeen, Hong Kong (China)

Fig. 11.37 From the fish market in Nice, France. The French sea food kitchen offers a great variety of alternatives

in the dry period. Ingenious irrigation systems developed from thousands of years often secure the watering of the rice fields. 90 % of the world's rice production and rice consumption takes place in Asia. Also, half of the world's production and consumption take place in the two most heavy populated countries in the world, China and India. Brazil that is number 10 on the ranking list is the only non-Asian country on the list of the largest rice producers in the world. Rice is an old food product. It is assumed that the first rice farming took place along the Pearl River in China for more than 10 000–15 000 years ago. Some sources also mention cultures along Yangtze River and on the Korean peninsula as the possible origin for rice production. To Europe the rice plant probably came from China via Egypt (Figs. 11.47, 11.48 and 11.49).

A Simple, "Quick" Fish-Dish

- A European touch based on a South East Asian basic idea (Fig. 11.50).

- Seafood—in this case some mussels, a homemade fish ball from a previous dinner, and some shrimps (it could instead, or in addition, been some filet of salmon, cod, or other sea food)

Fig. 11.38 The fish market in Bergen, Norway a rainy day in June. The offer is wide and the sitting space for the tasters are full, probably mostly by tourist, and very often of Asian origin

- 2–4 garlic of the mild type that can be cut in thin slices.

- 2–4 spring onions in reasonable pieces.

- half of a sweet paprika, that is cut in thin slices.

- 2–3 cm ginger root in thin slices.

- 2–3 slices from a box of pineapple that is cut up in pieces. (This gives the good sweet taste). Include some of the juice. A fresh pineapple is just as good, or even better.

- 3–4 table spoons from a box of cut and skinned tomatoes (Some skinned fresh tomatoes are even better, but it takes more time).

- a few table spoons of coconut oil or milk

- 2 table spoons of soy sauce.

- some olive oil is already in the frying pan.

First fry the thin slices of garlic in the pan until they has a golden color (Fig. 11.51).

The garlic flakes are taken out of the pan when they are golden, and are put on a separate plate. They leave a nice little garlic taste in the pan. The crisp garlic flakes are additions to the food, in the amount that is individually suited.

Turn the heat a bit down, and add the sea foods except the shrimps to the pan together with the ginger flakes and let them fry for a few minutes. Then add the rest of the ingredients. Put on slow heat for 5 min. Add the shrimps in the end (Fig. 11.52).

Meanwhile, the rice has been boiled. Everything is finished in 15–20 min.

Fig. 11.39 Fish market in Hua Hin, Thailand. Dried octopus in many variations. Typical South Asian countries have a larger offer in varieties of sea food than any other places in the world

The described portion should be suitable for two persons (Fig. 11.53).

An important factor for the production of the food we gets from water is of course the photosynthesis and an exciting and intricate ecosystem. The photosynthesis is the most important condition for all growth, but is less efficient in water due to the light refraction, and is reduced in degree of efficiency with water depth and distance from equator. In the coastal zones and on shallow waters the biomass production is greatest, and we find an intricate eco system from freely floating algae and freely driving plankton, a myriad of corals, small animals like sea stars and worms to shellfish and fish.

The animal life goes far deeper, and is recorded even at the greatest ocean depths, but it decreases considerably in number of species with the water depth.

The variation in these systems is not only due to temperature and sunlight, but also the coastal and ocean currents, waves, nutritious salts, and oxygen additions. One factor for the number of species is also the fact that ocean and coastal areas which never have been covered by ice through an ice age have had more time to develop the wide variety of corals, plants, and fish.

Human activity has in some cases changed or destroyed the balance in the ecosystems, as for example:

- Sea weed deserts, probably mostly caused by very one-sided catching conditions, where for example sea urchins have been the conquerors of the sea bottom, and which again have eaten down the sea weed forest. We find examples in Northern Norway, the west coast of Canada and in more southerly waters. Even the Great Barrier Reef has been reduced by the increased sea urchin growth.
- Slamming of fiords and bays. We find a number of examples of this in various countries, where the sea bottom life has been considerably changed. Unclean water from human activity has been transported into the sea, where silting and more dangerous chemical deposits have been slamming up the sea bottom. However, in many places in the world, this has been recognized and repair and rehabilitation work is going on.
- Coral death or so-called "whitening" has taken place in many places of the world, and probably more than any other places in the South China Sea, the Indian Ocean, and around Australia. The phenomena are still studied by

Fig. 11.40 Fish market Hua Hin, Thailand, the shell fish department. Every department in this little coastal town has offers in number of varieties. In a typical European coastal town there are many such departments

researchers, and final conclusions are still not drawn. But, we know something, which is to a large extent due to increase in temperature in the sea beyond the tolerance limit of some of the coral types. The consequence is that the corals die and get white. It will take tens of years before we can be sure of the final answer. In some areas the whitening does not have a human cause, but is due to earthquakes. Large areas of white corals have been tipped over like what might happen in a forest area after a heavy storm.

The many-fold in the oceans is dependent on that we find at muddy bottom, sandy bottom, and hard bottom as well as

A Simple, "Quick" Fish-Dish

Fig. 11.41 From the fish market in Tokyo, very orderly and delicate, but the prices might sometimes be relatively high

Fig. 11.42 The delicate counter of a Norwegian good local fish shop might easily tempt to at least one sea food dinner

Fig. 11.43 Delicate exhibition of tonight's offer of sea food outside a fish restaurant in Thailand. The offers in the menu of the restaurant are not less tempting

Fig. 11.44 Crab in yellow curry might look a bit strange, but the taste is fantastic

grottos and caves. Different species have different preferences for their locations. Some species even change their preferences during their life. An example is —that swims in the water in some times in its premature phase, then likes to dig hiding places in the sand, and as a mature creatures prefers caves to hide in.

If either the food or the preference bottom disappears for a species, it will be decimated or disappear.

A Simple, "Quick" Fish-Dish

These relationships are of uttermost importance if we shall have the possibility to get the food we need from the sea in the future.

For partly repairing the damages from human activity and partly utilizing the functions from the nature better, we have the establishment of artificial fish reefs (AFR). This has been increased considerably in the last tens of years.

Artificial Fish Reefs

The origin of the human created fish reefs was to increase the catches of fish. Later, purposely designed artificial fish reefs, or AFRs have been used to solve other environmental challenges. Some people have reacted negatively on the word "artificial," but these structures have over the last few tens of years got considerable use on most continents.

In 1990, the French researcher Simard made an investigation in Japan FCA (Fishery Cooperative Association) regarding the use of AFRs:

Fig. 11.45 The fish might be served in many ways

Fig. 11.46 The Thai kitchen has numerous sea food delicatessen. To the far: Ghoong Hom sabai—tiger prawns in an egg net; Nearest: A national favorite dish—Tom Yam Ghoong Nang—soup with shrimps, lemon grass, mushrooms etc

Fig. 11.47 Typical Japanese rice fields. Picture taken from the train ride between Tokyo and Kyoto

- Thirty eight of 40 answered that they used the reefs often.
- 77 % answered that the reefs increased the global catches.
- 77 % answered that they believed that the reef contributed to attract fish.
- 55 % answered that they believed that the reefs increased the reproduction.
- 37 % said that the reef increased the regularity of the catches.
- 32 % said that the reef increased the size of the fish.
- 15 % said that the reefs contributed to the quality of the catching.

Knowledge and experience around AFRs have developed to an important research topic that includes integrated expertise from a number of research and technology areas.

Some background history

We know there were traditions among Norwegian farmers well over hundred years ago that when cutting timber in the winter to leave branches on the frozen lakes. That was always a good fishing place the next summer.

The oldest documentation we know regarding AFRs is from Japan in 1795. Fishermen from the island of Awaji outside Kobe reported about good catches at a sunken ship. After 7–8 years, the ship disintegrated and the fish disappeared. The fishermen filled a timber raft with stone and sunk it on the depth of 40 m. Three-months later they got more fish on this location than on the earlier sunken ship [28].

A stone memorial in the town of Uoshima in Japan tells a similar story.

Over hundred years ago it was tradition of several places in Japan to sink old fishing boat to improve fishing. Similar traditions are also found several other places in the world. In 1920, the Japanese words jinco gyosho (fish reefs made by men) was used for the first time, and in 1923 the Japanese government provided the first subsidies for construction of AFR.

The first AFRs in concrete were built in Japan in 1954.

In the years 1988–1994 the Japanese government invested more than 1 billion USD per year in construction of AFRs in concrete [29].

China's experience in using AFR started from 1975 in

Fig. 11.48 Tamil Nadu, India. Rice harvest. The asphalt road is suitable for cleaning of the rice from the straw. The traffic just has to wait for a while

Beibu Wan (North Bay) of Guangxi Zhuang Autonomous Region and has not yet been progressing in a large and systematic volume.

Where have AFRs been used?

Seaman [30] refers to a bibliographic investigation in 1986 with reference to AFRs in 39 different countries.

Jahren made a *state of the art*—report in 1998 [30], and found references to AFRs in 61 countries.

AFRs have been utilized on all continents and in both salt water and fresh water. Japan has been a pioneering country in this field.

There are several possible motives behind the use of AFRs:
- To gather fish and thereby get a more cost efficient and sustainable fishing.
- Increase the efficiency of existing breeding area or build new ones to increase biomass production.
- Bring new life to sea bottom deserts
- Caused by coral death due to temperature changes in the water
- Caused by coral destruction from dynamite fishing
- Caused by slamming and silting
- Caused by kelp forest destruction.

Fig. 11.49 Rice Terrace, Longji, Guangxi Zhuang Autonomous Region, China. It is not only famous for the rice production, but also for tourists due to the beautiful scenery. October, 2010

Fig. 11.50 It only takes 5 min to prepare the ingredients

Artificial Fish Reefs

Fig. 11.51 The garlic slices are finished and can be removed from the pan

Fig. 11.52 The dish is ready for serving

- Purify/improve water quality to increase the many fold of marine life.
- Utilize waste nutrition for increased fish life and clean under fish farms.
- Refine the sea bottom for increasing growth of the most valuable sea food.

- Establish predictable fishing ground for tourists and thereby increase the season for beach hotels
- Secure fishing ground for the coastal population.
- Control beach erosion.

The most important thing to bear in mind with this list is that the reef design should take into consideration the motive for the establishment of the AFR. Unfortunately we see example of reefs established only to get rid of wastes without too little focus on the optimum design.

In 1984, USA approved the *National Fishing Enhancement Act*. This dHeavy growthocument gives a good general summary for the motivation of establishing AFR's;

> ...properly designed, constructed and located artificial reefs ... can enhance the habitat and diversity of fishery resources; enhance United States recreational and commercial fishing opportunities; increase the production of fishery products in the United States; increase the energy efficiency of recreational and commercial fisheries; and contribute to the United States and coastal economics.

The size of the reef and the number of reef modules are important factors in designing AFRs. Researchers claim that the reef should be at least 20 000 m^2 for a new AFRs area, or at least 5 000 m^2 for extending an existing fishing ground. Other researchers claim that such numbers have to be modified depending on the location and types of fish.

The Chinese fish industry has over time grown to be the biggest in the world in consumption, farming, catching, import and export. China has a long coast with farming not only of fish, but also other delicatessen on the wide Chinese menu.

In a speech during a seminar in Shenyang in China, October 2008 about sustainable development of the concrete industry, Professor Ming Tang reported about advanced testing of concrete modules for farming of abalone and sea cucumbers (mentioned earlier in this chapter) [31]. Altogether 10 000 units of the abalone modules alone were produced (Fig. 11.54).

Observations showed growth on the modules already 20 days after they were set out into the sea, and colonization of abalone after 40 days.

The first Norwegian AFRs were placed in the sea outside the island of Vestvågøy in Lofoten in Northern Norway in 1994. The reefs were placed to places on 20 and 40 m depth, respectively [32] (Fig. 11.55).

Fig. 11.53 Good appetite—South Asian seafood in a Norwegian setting

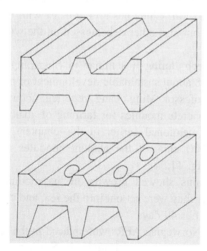

Fig. 11.54 Abalone reef modules

Already in 1995 growing on the reef modules were observed, and the amount of fish had increased 4–5 times compared to a reference area.

In October 1994, 10 years after the placing of the reef modules, marine biologist from the Norwegian Institute of Water investigated the reefs by diving and ROV. They concluded that the reefs were in good shape and had a very positive effect on plants and fish life.

Tjuvholmen

The competition about the design and construction of the Tjuvholmen area in Oslo harbor was held in 2002. In the winner design included rehabilitation of the sea bottom including use of artificial fish reefs. This was also done when the work started in 2005.

The motive behind the AFR solution was twofold:

- Cleaner water.
- More maritime life.

The solution consisted of two main actions with five different types of components. Before start the bottom was a sea bottom desert with muddy bottom. The start was therefore to position a layer of 40 cm of sand on the bottom.

The water improvement tools were established under the quay deck between the parking structures.

Fig. 11.55 One of the reef modules used in the first Norwegian AFR's —JAFI 2000. Here shown at the Lofot museum in Kabelvåg, Photo: Elisabeth Jahren

More than 1000 8–10 m long mussel ropes were hung up at intervals of 2 × 2 m under the deck (Fig. 11.56).

Blue mussel has a miraculous way to filtrate water. A mature mussel filtrates 4–6 L of water every hour. With 10 mussels for each 10 cm of the mussel rope, which means that the whole water table under Tjuvholmen is filtrated about every 10 h. The water depth varies from about 9–15 m.

At the bottom and under the mussel ropes are placed about 400 *cleaning reef modules*. These modules consist of a concrete plate with thread elements that create a very large area right above the bottom. These modules have a threefold function (Fig. 11.57):

- Be home for smaller organisms that have a supplementary cleaning effect near the bottom.
- Be home for fish and organisms that eat the old mussels that fall down from the ropes, and thereby prevent an unhealthy environment on the bottom.
- Contribute to a heterogeneous bottom environment.

On the sea bottom, along the quay front are placed 20 reef groups, each with 10 modules of different shapes. These modules both provide shady hiding places, caves, channels, and swim through holes for fishes that live above the sea bottom (Figs. 11.58 and 11.59).

Fig. 11.56 Mussel ropes

Fig. 11.57 Cleaning reef modules after half a year on the bottom (*left*) and before setting out (*right*)

Fig. 11.58 Model of the reef positioning

Marine biologists made investigations in 2009 and 2010, and amongst others concluded that (Fig. 11.60):

- There was heavy growth (more than predicted) on the blue mussel ropes
- The reefs function well.
- Heavy growth and marine life in the "cleaning reef" modules were found.
- The animal life seems to stabilize.
- Improvement in the water quality was also found.

Fig. 11.59 Reef modules in the harbor, ready for placing in the sea

Fig. 11.60 Maritime life in the Tjuvhiolmen reefs after 1 year

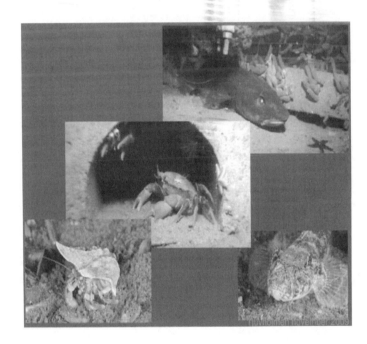

References

1. Jahren Per: *Kunstige fiskerev,* Rapport 01.10.98, P.J.Consult AS for NFR, Oslo.
2. FAO: *FAO Yearbok 2002,* FAO 2005.
3. Muus Bent J.med tegninger av Dahlstrøm Preben: *Våre saltvannsfisker,* NKS-Forlaget, Oslo, Norway, 1981.
4. WWDR4: *Facts and Figures—Managing Water under Uncertainty and Risk,* From the United Nations Water Development Report 4. Internet 16.09.12.
5. Statistisk Sentralbyrå (Statistical Central Bureau): *Statistiske analyser(Statistical analysis)—Naturresurser og miljø (Natural resourses and environment) 2005,* Statistisk Sentralbyrå, Oslo-Kongsvinger, Norway, December 2005.
6. Statistisk Sentralbyrå (Central Statistical Bureau): *Statistisk årbok (Statistical Yearbook) 2011,* SSB, Oslo/Kongsvinger, Norway, August 2011.
7. Kirkebøen Stein Erik: *En verden under havet,* Aftenposten, Oslo, Norway, 10.09.2012.
8. Internet 24.12.12: http://www.mollers.no/c-51-Historie.aspx.
9. Internet 27.12.12: *Klippfisk* http://no.wikipedia.org/wiki/Klippfisk.

10. Internet 06.02.13: *Hvalfangst*, http://no.wikipedia.org/wiki/Hvalfangst.
11. Internet 08.02.13: *Svend Foyn*, http://no.wikipedia.org/wiki/Svend_Foyn.
12. Internet 07.02.13: *Norsk hvalfangst*, http://www.hvalfangst.info/informasjon/fangsten/norsk-hvalfangst/.
13. Jacobsen Hanne Østli: *Foreslår kvotesystem for hvalfangst*, http://www.forskning.no/artikler/2012/januar/310303.
14. World Watch Institute: *State of the World 2012*. Island press, Washington DC, USA 2012.
15. Hamnes Lena Amalie: *Naken uten fisk.(Naked without fish)* H. Aschehoug & Co, Oslo, Norway, 2006.
16. Odd Richard Valmot: *Fiskefor for fremtiden*, Teknisk Ukeblad, Oslo, Norway. 11/13.
17. Kunnskapsforlaget/Norges Naturvernforbund: *Natur—og miljøleksikon*, Hovedredaktør: Henning Even Larsen, Oslo, Norway, 1991.
18. Internet 13.10.12. Wikipedia: *Coromant fishing*, http://e.wikipedia.org.wiki/coromant_fishing.
19. Enerstvedt Vidar: *Rådyrt! Kjøpte tunfisk til 10 millioner – ble servert som sushi*...VG. Oslo, Norway, 6. January 2013.
20. Internet 15.01.13: *Tuna*, http://en.wikipedia.org/wiki/Tuna.
21. Internet 03.03.13: *Frog legs*. http://en.wikipedia.org/wiki/Frog_legs.
22. Internet 05.09.12: Stortare.no, http://www.stortare.no/.
23. Pronova Biopolymer AS, Redaktør, Trond Aasland: *Utfordringen fra havet*, Pronova Biopolymer AS, Drammen, Norway,1997.
24. Internet 02.09.13: Wikipedia: Spirulina (dietary supplement) http://en.wikipedia.org/wiki/spirulina_(dietary_supplement).
25. Internet 18.03.13; *FAO Fisheries & Aquaculture Ostrea edulis*, http://www.fao.org/fishery/culturedspecies/ostrea-edulis/en.
26. Internet 01.01.13: *Krill*, http://en.wikipedia.or/wiki/Krill.
27. Internet 11.03.13: *Liste over norske fjorder (list of Norwegian Fiords)*. http://no.wikipedia.org/wiki/liste_over_norske fjorder.
28. Rivenga et al: *European Artificial reef Reseach*. Proceedings of the 1st EARRN Conference, Ancona, Italy, March 1996.
29. Jahren Per: *Japansk Fiskehusteknologi – orienterende undersøkelser*, P.J. Consult AS/Fiskehus AS, Asker, Norway, Report 10.04.90.
30. Seaman W jr and Sprague L.M: *Artificial Habitats for Marine and Freshwater Fisheries*. Academic Press Inc, San Diego, USA, 1990.
31. Ming T..*Research and High performance Artificial reef for Breeding Abalone and sea cucumber in Sea Water*, International workshop on sustainability in the cement and concrete industry, Shenyang, China, October 2008.
32. Jahren P. : *Artificial Reef project, Vestvågøy, Lofoten, Norway 1994—Background, Design and experience*. EARRN-meeting, Southampton University, UK, January 1998.
33. Internet 20.09.12: FAO, Yearbook of fishery statistics, ftp.fao.org/fi/stat/summary/default.htm.